Fun Games and Activities for Teaching Times Tables

Debbie Chalmers

We hope you and your pupils enjoy using the ideas in this book. Brilliant Publications publishes many other books to help primary school teachers. To find out more details on any of the titles listed below, please log on to our website: www.brilliantpublications.co.uk.

By the same author
Teaching Grammar, Punctuation and Spelling Through Drama
Speaking and Listening Activities for the Early Years
Drama Activities for the Early Years

Maths titles
The Mighty Multiples Times Table Challenge
Mighty Fun Activities for Practising Times Tables (3 book series)
Maths Problem Solving (6 book series)
Open-ending Maths Investigations (3 book series)
Deck Ahoy! Primary Mathematics Activities and Games Using Just a Deck of Cards
Sum Fun Maths Assessment (3 book series)
Maths Problems and Investigations (3 book series)

Published by Brilliant Publications
Unit 10
Sparrow Hall Farm
Edlesborough
Dunstable
Bedfordshire
LU6 2ES, UK

www.brilliantpublications.co.uk

The name Brilliant Publications and the logo are registered trademarks.

Written by Debbie Chalmers
Illustrated by Vicky Chalmers

© Text Debbie Chalmers 2017
© Design Brilliant Publications 2017

Print   ISBN 978-1-78317-274-0
ePDF   ISBN 978-1-78317-275-7

First printed and published in the UK in 2017

The right of Debbie Chalmers to be identified as the author of this work has been asserted by her in accordance with the Copyright, Designs and Patents Act 1988.

Pages 5–75 may be photocopied by individual teachers acting on behalf of the purchasing institution for classroom use only, without permission from the publisher and without declaration to the Publishers Licensing Society. The materials may not be reproduced in any other form or for any other purpose without the prior permission of the publisher.

Contents

Introduction 4	We all stand together 40
Number grid 5	In the ring 41
	Break it down 42
	Gone shopping 43
	Stand over there 44
	Keeping time 45

2 times table

| Ups and downs .... 6 |
| Groups of two .... 7 |
| Cross the line .... 8 |
| Hop and jump .... 9 |
| Find your number .... 10 |
| Around in a circle .... 11 |
| Score and collect .... 12 |
| Toss them in .... 13 |
| Record the sum .... 14 |
| Count them up .... 15 |

Learning in sequence

| Toss it over .... 46 |
| Make a list .... 47 |
| Turn and turn again .... 48 |
| Count and catch .... 49 |
| Into the middle .... 50 |
| How far can you go? .... 51 |
| Take your partners .... 52 |
| Every increasing circles .... 53 |
| Stand in line .... 54 |
| A few more to go .... 55 |

5 times table

| Mind the gap .... 16 |
| Into the bucket .... 17 |
| Along the line .... 18 |
| Bounce with me .... 19 |
| Take a step forward .... 20 |
| Shake it up .... 21 |
| How many minutes? .... 22 |
| We go together .... 23 |
| Inner circles .... 24 |
| Pile it on .... 25 |

Guessing the table

| A model experience .... 56 |
| How many in the group? .... 57 |
| Pick me up .... 58 |
| Find your partners .... 59 |
| Show me your answers .... 60 |
| One, two, three .... 61 |
| Look underneath .... 62 |
| Just choose one .... 63 |
| Make a record .... 64 |
| Follow the trail .... 65 |

10 times table

| Towering heights .... 26 |
| Jump and turn .... 27 |
| What's the catch? .... 28 |
| Count up and count down .... 29 |
| Meet in the middle .... 30 |
| Magic rings .... 31 |
| Throw it aside .... 32 |
| Catch this! .... 33 |
| Roll in order .... 34 |
| Going round in circles .... 35 |

Finding factors

| Swap with me .... 66 |
| Factor that in .... 67 |
| Either way .... 68 |
| It works both ways .... 69 |
| Put them together .... 70 |
| Two bounces .... 71 |
| Throw them to us .... 72 |
| Walking two by two .... 73 |
| Find a partner .... 74 |
| Set it out .... 75 |

2, 5 and 10 times tables

| Build it up .... 36 |
| Knock them down .... 37 |
| Change the shape .... 38 |
| Pass it on .... 39 |

Introduction

As it is now a requirement that all children leaving primary school must have a sound grasp of the 2–12 times tables, teachers and others who work with them must look for stimulating and imaginative ways to teach these tables and try to make the learning process both effective and fun.

The first chapters in the book contain activities specifically for use with the 2, 5 and 10 times tables, first on their own, and then together.

The remainder of the games are appropriate for any of the times tables, from 2–12, and may be used to practise whichever tables are being learned by the children in the class or group at the time of playing.

Some of the games offer opportunities for children to learn the questions and answers for each table in sequence. Other games test their memories and thinking skills as they try to identify the table they are working on. There are also games to teach children the different factors that can make up each answer.

When games require children to take on different roles, you should choose the parts carefully to suit their individual abilities and confidence levels, ensuring that appropriate challenges are offered to each. The degree of stimulation and challenge may be increased gradually to suit each class or group, by moving from simpler tables to harder ones, swapping from sequences to random orders or working at a faster pace.

Children learn best when they are presented with information repeatedly and in a variety of ways. The games in this book help to teach times tables through a mixture of visual, aural and kinaesthetic activities: looking and sorting number cards, saying the numbers aloud and moving around to find the correct places and partners.

To prepare for the games, you will need to make sets of cards, each large and clear enough to be seen from a distance. These may be as simple as sheets of paper or, to make them last longer, squares of plain card/laminated sheets. The numerals can be written with a black marker pen, and must be large enough to be instantly recognisable and easy to distinguish from one another from a short distance away. Children may make the number cards for themselves, but you should check to ensure that the numerals are correctly formed, of an equal size and large enough.

To be able to play any or all of the games, it is necessary to create two sets of cards numbered 1–12 and thirteen sets of cards numbered 0–9 that will be reusable on several occasions. If making sets of cards for single use only, the numerals needed for each game are listed under Preparation. Some activities will also require 'x' and '=' cards.

The games will be most successfully played in a hall or other large clear space, or outside on a playground or grassy area. However, they may be adapted to fit within any available room or outdoor space, depending upon the number of players. Most of the games will ideally be played with a class of 20–30 pupils, but, for some games, a smaller number can encourage more active participation, while other games are more fun when played with a larger group.

Number grid

The following grid may be useful when planning games. There may be times when it is necessary to quickly check exactly how many times a particular answer occurs within the tables, how many factors a multiple has or which number appears most frequently.

1	2	3	4	5	6	7	8	9	10	11	12
2	4	6	8	10	12	14	16	18	20	22	24
3	6	9	12	15	18	21	24	27	30	33	36
4	8	12	16	20	24	28	32	36	40	44	48
5	10	15	20	25	30	35	40	45	50	55	60
6	12	18	24	30	36	42	48	54	60	66	72
7	14	21	28	35	42	49	56	63	70	77	84
8	16	24	32	40	48	56	64	72	80	88	96
9	18	27	36	45	54	63	72	81	90	99	108
10	20	30	40	50	60	70	80	90	100	110	120
11	22	33	44	55	66	77	88	99	110	121	132
12	24	36	48	60	72	84	96	108	120	132	144

It is possible to see from this grid that using two sets of cards 1–12 and 13 sets of cards 0–9 will enable a class or group to create each one- and two-digit number needed to ask the number questions and all one-, two- and three-digit numbers that occur as answers within the times tables 2–12.

Numbers 12, 18, 20, 30, 40, 48, 60 and 72 each occur four times within the 2–12 times tables; 36 occurs five times and 24 is the most frequent multiple, occurring six times.

Nine multiples have two sets of factors: 12, 16, 18, 20, 30, 40, 48, 60 and 72. There are two multiples that have three sets of factors: 24 and 36.

Ups and downs

2 times table

Activity

Ask children to sit in a line across the floor, then to turn to the person beside them and hold hands to form partners for the game. Ask the children to shuffle a little to give each pair a separate space.

Call out the multiples that form the answers within the 2 times table, one number at a time. For example:

| 2 | 4 | 6 | 8 | 10 | and so on to | 24 |

Starting from the left of the line, the first pair of children stands up as they hear 2; the second pair of children stand up on 4, the third two children on 6 and so on to 24.

Then call out the whole multiplication statements in the same way. For example:

| 1 x 2 = 2 | 2 x 2 = 4 | 3 x 2 = 6 | 4 x 2 = 8 | 5 x 2 = 10 | and so on to | 12 x 2 = 24 |

This time each pair sits back down when they hear their statement.

Extension/challenge

Ask children to say the numbers and the multiplication statements for themselves (with help from you or each other when necessary), as they stand up and sit down in the correct order.

Learning objective
Understanding that counting in 2s in ascending order creates the 2 times table.

Preparation
No equipment needed.

Number of players
24. If more, you can have more than one pair of children for some multiples (with the second pair sitting behind the first and playing simultaneously with the pair in front).

Groups of two

2 times table

Activity
Count together how many pairs of pupils there are.

Call out a number that is a multiple answer in the 2 times table. Pairs of children then join together to make a group of that number and discuss together how many twos are in the group.

For example, if 4 is called, sets of two pairs join together and each group discusses 2 x 2 = 4.

If 12 is called, sets of six pairs join together and each group discusses 6 x 2 = 12.

Learning objective
Understanding that thinking of pairs and counting in 2s in ascending order creates the 2 times table.

Preparation
No equipment needed.

Number of players
24+, standing in pairs around the space.

When the appropriate number has been agreed and the multiplication statement chanted aloud, the pairs separate again and wait for the next number to be called.

The number of pairs will not always divide evenly within the group of players. When this happens, the first children to form the groups discuss the multiplication statements and chant them aloud, while the others watch and learn from them until the next number is called. The chart below shows which multiples will divide equally and which won't for a class of 24. (You can adapt it to suit your class size.)

Divides evenly with class of 24	Won't divide evenly with class of 24
1 x 2	5 x 2
2 x 2	7 x 2
3 x 2	8 x 2
4 x 2	9 x 2
6 x 2	10 x 2
12 x 2	11 x 2

Extension/challenge
Ask children to chant the 2 times table multiples for themselves, in correct ascending order, and to form groups and chant the multiplication statements aloud for each one, working as a whole class.

Cross the line

Activity
Ask 12 children to stand behind the cards 1–12. Ask 6 children to stand behind the cards 14, 16, 18, 20, 22 and 24, leaving empty spaces behind the odd numbers 15–23.

Stand in the centre, holding the ball and facing the numbers 1–12, with the higher numbers behind them. While throwing the ball to child number 1, chant 1 x 2 = __ . As the children shout 2, child number 1 throws the ball to child number 2 and child number 2 then throws the ball back to you.

While throwing the ball to child number 2, chants 2 x 2 = __ . As the children shout 4, child number 2 throws the ball to child number 4 and child number 4 then throws the ball back to you. The game continues in this way with child number 3, then 4 and so on. When child number 7 is reached, they will need to throw the ball across to number 14 in the other line and the game continues, involving the higher numbers until you reach 12 x 2 = 24.

Learning objective
Reinforcing the 2 times table by chanting it aloud, rhythmically, in ascending order.

Preparation
- Set of number cards 1–24. Lay out number cards 1–12 in a line beside each other and number cards 13–24 in a line facing them
- Large ball (foam if playing inside).

Number of players
Ideally, play this game with 18 children (or create two simultaneous games with 36 players, involving adults if necessary). If a different number of children need to play, allow some numbers to be represented by pairs of children instead of single players.

Extension/challenge
Play the game the other way around, with you chanting the answers and the children choosing the factor that combines with 2 to make the multiple. For example: 24 = 2 x __ ;
22 = 2 x __ .

The child standing behind the multiple number must choose the factor number to throw the ball to, helped by the rest of the class calling out the answer. For example: child number 24 will throw the ball to child number 12 and child number 22 will throw the ball to child number 11 and so on.

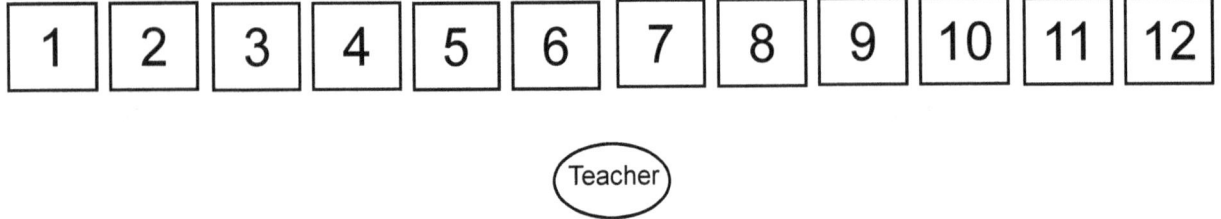

Fun Games and Activities for Teaching Times Tables © Debbie Chalmers and Brilliant Publications
This page may be photocopied for use by the purchasing institution only.

Hop and jump

2 times table

Activity
Ask children to take turns to hop and jump (or jump with two feet together and then two feet separately) into the hoops in turn from one end to the other and back again. When they reach number 7 (in the 10th hoop), they need to jump around to face the other way and return to the beginning. As they jump, they should stay silent for the odd numbers (thinking of them in their heads) and chant aloud the even numbers: 2, 4, 6, 8, 10, 12.

After each child has had a turn, the children can take another turn, this time using the hoops twice, to go to the end and back and then to the end again and back again, to chant the even numbers to 24, completing the whole 2 times table.

Extension/challenge
Remove the single hoops and leave only the pairs laid out. Invite children to jump through them quickly, chanting the even numbers only without needing to think of the odd numbers in between.

Learning objective
Practising counting in 2s and chanting the multiples of the 2 times table aloud in ascending order.

Preparation
Lay out ten hoops in a typical hopscotch pattern:

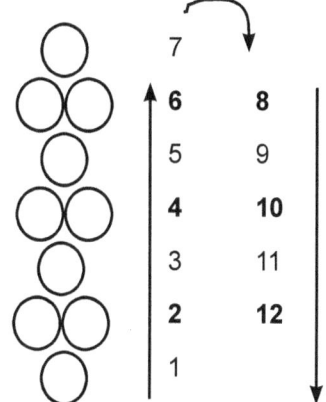

(Alternatively, chalk circles or squares onto an outdoor surface.)

Number of players
Any number. Set up extra rows of hoops to allow more than one child to hop at the same time and avoid children having to wait too long for their turns.

© Debbie Chalmers and Brilliant Publications — Fun Games and Activities for Teaching Times Tables

Find your number

2 times table

Activity
Ask each child in turn to step or jump to the correct answers while the group chants:

| 1 x 2 = 2 | 2 x 2 = 4 | 3 x 2 = 6 | and so on to | 12 x 2 = 24 |

They should notice that they are only landing on the even numbers.

Repeat with each child taking a turn to step or jump along the line while the group chants the multiples only:

| 2 | 4 | 6 | and so on to | 24 |

Learning objective
Understanding and reinforcing the ascending order and rhythm of the multiples of the 2 times table.

Preparation
Number line 1–24 with each number in a square, large enough for a child to stand on. (Either attach paper or card squares to the ground or chalk squares onto an outdoor surface.)

Number of players
Any number from 3 to 12.

Extension/challenge
Ask children to stand in a line so that they know whose turn is next. Call out multiplication questions from the 2 times table in a random order. For example:

| 11 x 2 | 10 x 2 | 3 x 2 | 12 x 2 | 2 x 2 |

As each statement is called, the next child in line steps or jumps onto the correct product in the number line. For example:

| 22 | 20 | 6 | 24 | 4 |

You must be careful not to repeat a statement, so that the correct square is always empty and available. This game should be played either with 6 children (two turns each per round of the game) or 12 children (one turn each per round). It is more fun with 12, as the children like to fill each square and end the game by standing very close together, all balancing on the number line. Any number of rounds can be played, with all of the children jumping off the number line after 12 turns have been taken and forming a new line some distance away, ready to start again.

Fun Games and Activities for Teaching Times Tables © Debbie Chalmers and Brilliant Publications
This page may be photocopied for use by the purchasing institution only.

Around in a circle

2 times table

Activity
Ask 12 children to stand behind the number cards in the circle.

Stand in the centre and call out a multiplication question from the 2 times table in a random order and then throw the ball to the correct player (alternatively, a child could do this, with your assistance). The child standing by the correct product must call out their number and catch the ball as it is thrown to them and then throw it back.

Learning objective
Remembering and practising chanting the sums of the 2 times table aloud in random order.

Preparation
- Set of 12 even numbered cards 2–24, placed in a circle in clockwise order on the ground
- Large ball (foam if playing inside).

Number of players
12 (ideally). If you have 24 or 36 players, make two or three circles with two or three adults and play simultaneously. If there are more or fewer players, allow some children to stand in the centre of the circles and call out the sums, helped by an adult if necessary. (Lists to read from could be prepared in advance.)

Each number should be called at least twice, but some more often. Take care to call the numbers in a completely random order, or some children may work out which numbers are left and stop listening and learning once they have had their turn.

Extension/challenge
Invite players to swap places each time a number is called. For example: the centre player calls 4 x 2 = __ ; child number 8 answers and catches the ball and then they swap places.

© Debbie Chalmers and Brilliant Publications   Fun Games and Activities for Teaching Times Tables

Score and collect

2 times table

Activity

Ask children to take turns to roll the two balls and knock down as many skittles as possible. Then they should count up the total that they have knocked down and take the factor card that is linked to that multiple within the 2 times table. For example: knock down 24 skittles and take the 12 card; knock down 12 skittles and take the 6 card.

Totals can be split to enable more cards to be taken. For example: if 20 skittles are knocked down and the 10 card has already been taken, then two cards that add up to 10 may be taken. The aim is to eventually take all of the cards.

If an odd number of skittles is knocked down, or all totals for the factor have already been used, the skittles must be replaced and that turn taken again.

Learning objective
Linking factors and multiples within the 2 times table.

Preparation
- Number cards 1–12, laid face up
- 24 skittles, set up in random pattern
- 2 suitable balls.

Number of players
Any number up to 12 may take part, but it may be better to play more than one game simultaneously (if enough skittles are available), to avoid children having to wait too long for a turn.

Children play together, taking turns to roll the balls, choose the cards, discuss and negotiate, aiming to collect all the cards as a group. You could suggest that they try to complete the activity as quickly as possible, or to beat their previous time, to make the game more exciting and appealing.

Extension/challenge

Set up the skittles in a random pattern. Shuffle the cards and place them in a pile, face down. Ask children to turn over the cards one at a time. For each card, they need to work out the correct multiple for that factor (that number of 2s) and roll the balls to knock down the correct number of skittles. For example: turn over the 6 card and knock down 12 skittles.

If they manage to knock down the correct number of skittles, they may keep the card and move on to the next one. If they accidentally knock down too many skittles, they should replace the card at the bottom of the pile to be attempted again later and take the next card from the top of the pile. Children work as a whole group, aiming to eventually collect all the cards together.

Toss them in

2 times table

Activity
Children may work alone or in pairs. Ask each child or pair in turn to pick up and turn over the top card from the pile and throw that number of beanbags into the first bucket. They must then work out the sum and throw the correct number of beanbags into the second bucket. For example: if five beanbags were thrown into the first bucket, then ten beanbags must be thrown into the second bucket

5 x 2 = 10

Extension/challenge
Let children challenge each other. One child or pair throws a number of beanbags into the first bucket and another child or pair works out the multiple and throws the correct number of beanbags into the second bucket. Then they swap places and repeat the activity the other way around.

Learning objective
Recognising how to record the different parts of the 2 times table.

Preparation
- Number cards 1–12, shuffled and placed face down in a pile
- Extra cards showing x, 2 and =, laid out on the ground with a bucket at either end of the number problem.
- 36 beanbags.

Number of players
Any number. If the group is small, children should play alone, but if the group is large, they should play in pairs. Encourage children to pay attention to the turns taken by the other players as there will be only one chance to form each sum each time the game is played.

Record the sum

2 times table

Activity

Choose two children to be the number 2 and ask them to hold a 2 card together and stand between the 'x' and '=' cards. Ask another child to turn over a number card from the pile and hold it up. For example 4. That child then walks around the space and chooses the correct number of children to join him/her to make a group that matches the card. For example, 3 more children to make a group of 4. They take the card and stand to the left of the 'x' card.

Other players then form pairs and join a pair chosen by you until the group agrees that there are the right number of twos (for example 4 pairs). The whole class chants together to count the children in the product pairs and decide on the right answer. For example 8. That many children then form a group, find the correct card and stand to the right of the = card.

The whole class should then be able to see the sum and say it aloud. For example: 4 x 2 = 8.

Extension/challenge

Place the two children (with the number 2 card) sometimes at the beginning of the sum and sometimes in the middle. Help the class to see that the answer will be the same whichever way the digits are multiplied.

Learning objective
Visualising the numbers and mathematical symbols used to record the 2 times table.

Preparation
- Number cards 2–12, shuffled and placed face down in a pile
- Even cards 4–24
- Extra cards showing x, 2 and = , laid out on the ground with a space at either end of the number problem.

Number of players
Minimum 28 (more if possible).

Count them up

2 times table

Activity

Ask each group to work together to put their coins into piles of one coin, two coins, three coins and so on up to twelve coins.

When they have done this, ask them to tip over one pile at a time and count the number of 2s (each coin is a '2') and then use their knowledge of counting in twos to decide the total that this makes. For example:

   1 x 2 = 2

   2 x 2 = 4

     3 x 2 = 6

Learning objective
Using knowledge of the 2 times table when counting coins and working with money.

Preparation
2p coins (78 coins per group) – either real coins (bags of coins are available from banks), play money or rubbings of 2p coins stuck onto circles of cardboard. (If using plastic coins or rubbings, provide a few real coins for the children to examine and handle first to ensure that they will recognise them in real situations.)

Number of players
The number of coins or rubbings that can be obtained or made will determine the number of players, but aim for 3–6 in a group and either 1 or 2 groups with an adult. If a large quantity of coins or rubbings is available, it is possible for more groups to play simultaneously with different adults.

Extension/challenge

Call out multiples from the 2 times table in a random order for the group and the children quickly decide how many 2s are in each multiple and present that number of coins.

Mind the gap

5 times table

Activity
Ask 30 children to stand in a long line, with gaps between them.

Invite two children to stand, one each side of the line, a short distance away and roll the hoop to each other. They should move up the line, rolling it through the gaps between each group of five children and chanting: 5, 10, 15, 20, 25, 30. Then they should move back down the line, rolling the hoop through the gaps between each group of five children again and chanting: 35, 40, 45, 50, 55, 60.

If preferred, children can swap places and two different children can take a turn at rolling the hoop through the gaps as they move back down the line. The game should continue until each player has taken a turn. All players can help by calling out where the right gaps are and chanting the numbers in order.

Learning objective
Understanding groups of 5 and how they can be put together to create the 5 times table.

Preparation
Large hoop that can be easily rolled along the ground travelling some distance before it falls over.

Number of players
32. Adults may join in with children if the group is not large enough.

Extension/challenge
Call out multiple numbers at random while two children roll the hoop through the right gaps, with the help of the other players calling out instructions and advice.

Try this first with the line representing 5–30, then try it with the line representing 35–60. For example: 10, 20, 15, 30, 5, 25, then 50, 40, 45, 35, 60, 55.

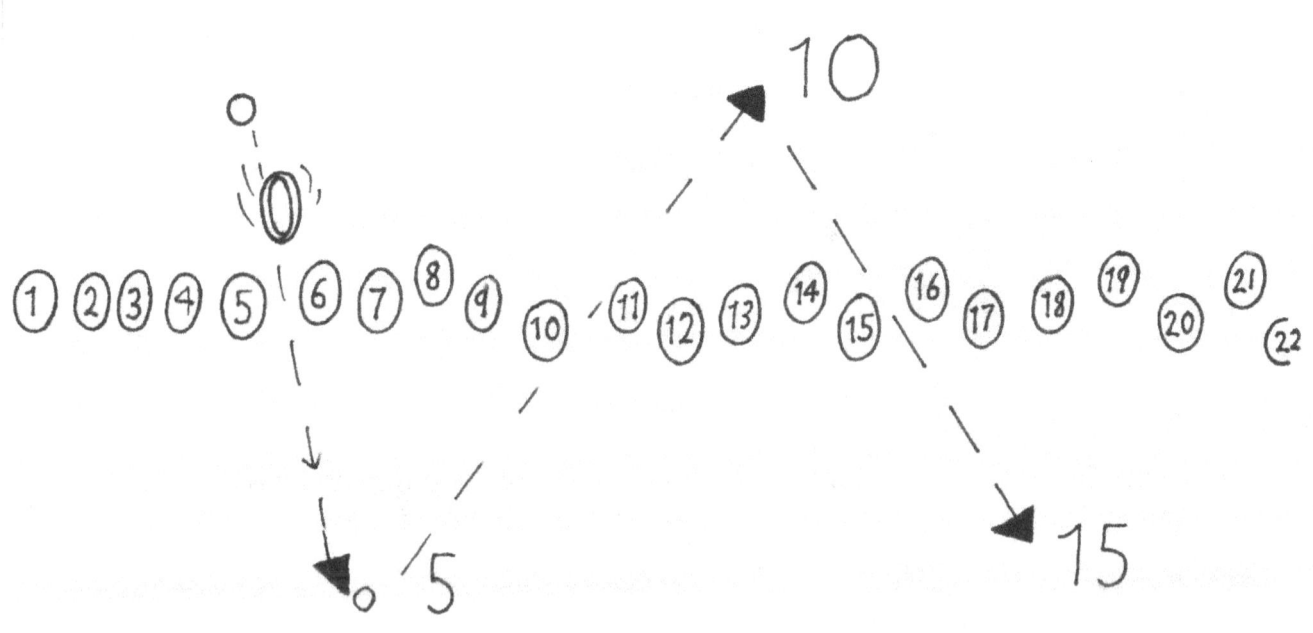

Fun Games and Activities for Teaching Times Tables © Debbie Chalmers and Brilliant Publications

Into the bucket

5 times table

Activity
One child at a time takes one step forward and throws a beanbag into container number 5, saying: 1 x 5 = 5. Then they walk back to the line, turn around, take two steps forward and throw a beanbag into container number 10, saying: 2 x 5 = 10. They continue to do this until they have taken 12 steps and thrown a beanbag into container number 60, saying: 12 x 5 = 60. All the beanbags are then gathered up and returned to the start line for the next player.

Depending on the number of players, each child can throw all twelve beanbags, or children can form teams and take turns to throw two or more each, saying the sums aloud together.

Extension/challenge
Ask children to try throwing the beanbags into the containers and saying the sums aloud in turn as before, but without walking forward the correct number of steps. This tests both throwing skills and memories and can give children a clearer idea of the multiples becoming larger as they multiply by an ever increasing number of 5s.

Learning objective
Understanding that counting on in 5s in ascending order, to make larger and larger numbers, creates the 5 times table.

Preparation
- 12 small beanbags
- 12 plastic buckets, toy boxes or crates labelled with multiples of 5 from 5–60. Lay the containers out in a diagonal line across the space, so that the numbers can be seen and each container is slightly further away than the one with the previous number. Make a line on the floor, a short distance back from container number 5, behind which players will stand to start the game.

Number of players
Any number.

Along the line

5 times table

Activity

Ask children to form groups of five and stand in separate groups around the space. Count together how many groups of five there are.

The players of the first group stand together at one side of the space, count themselves and call out 1 x 5 = 5. The second group stands beside them and they count all the players and call out 2 x 5 = 10. The third, fourth, fifth and sixth groups repeat this in turn, counting themselves and calling out each multiplication sentence. Encourage all children to count on 5 from the previous number, not start counting from 1 again.

Learning objective
Understanding that counting on in 5s in ascending order creates the 5 times table.

Preparation
No equipment needed.

Number of players
Ideally class of 30, so that 6 groups of 5 may be formed. If there are fewer, adults can play to make up the numbers.

When the children reach 6 x 5 = 30, the first group moves from the beginning of the line to the end and stands beside the sixth group to count on from 30 to 35. The class calls out 7 x 5 = 35. Each group in turn moves to the other end in this way until 12 x 5 = 60 is reached.

Extension/challenge

Ask each group of children to remember the two multiples that their group called out. For example: 5 and 35, 15 and 40 or 25 and 55. Chant the multiples in correct ascending order with the whole class (5–60), inviting each group to speak in turn and provide their own numbers on cue.

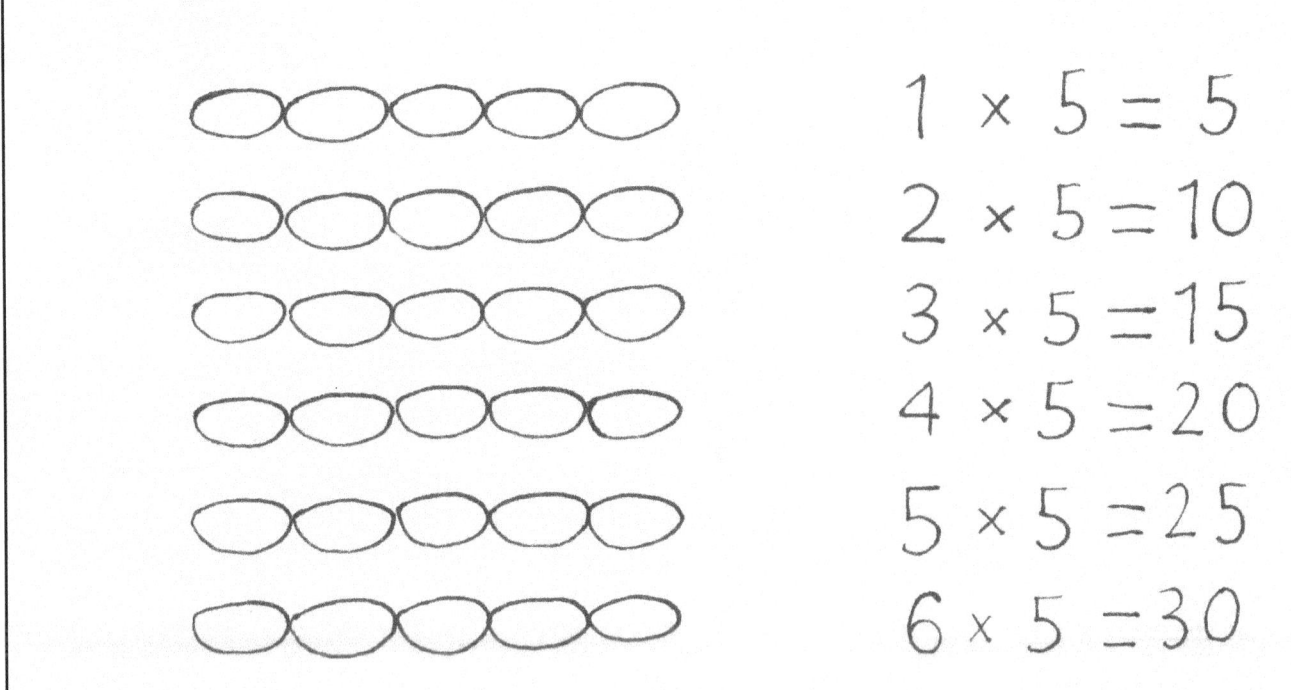

Fun Games and Activities for Teaching Times Tables © Debbie Chalmers and Brilliant Publications

Bounce with me

5 times table

Activity
Ask one child to bounce the ball 5 times, counting the bounces aloud, and then say the first multiplication sentence to the class: 1 x 5 = 5. Ask two other children to bounce the ball 5 times each, one after the other, counting the bounces aloud and then say the second sentence to the class: 2 x 5 = 10.

Different children continue in this way until there are no more or not enough left to form the next group. When each child has played at least once, they separate and re-shuffle to create groups of 8, 9 and 10 (which involves 27 children in total), and groups of 11 and 12 (23 children in total). Any who are not involved in a group at any time may help with the counting aloud of the bounces.

Extension/challenge
Secretly give each child a number 1–12 and ask them in turn to work out the total of their number x5 and to bounce the ball that number of times. Match this to ability, giving the more able and confident children the higher numbers and those who are still developing skills the lower numbers. Ask the class to count the bounces and guess the numbers.

Learning objective
Remembering and practising counting on to find the multiples in ascending order to create the 5 times table.

Preparation
Large bouncy ball (soft if playing inside).

Number of players
A group of 21 or 28 children is ideal for this game, as they can form six or seven groups exactly, involving each child once. If there are fewer, adults may play too to make up the number required or some children can participate in more than one small group.

© Debbie Chalmers and Brilliant Publications — Fun Games and Activities for Teaching Times Tables

Take a step forward

5 times table

Activity
Give 12 children one card each. The children must look at their own and each other's cards and put themselves into the correct order and stand in a line.

The child holding the number 5 card says 1 x 5, walks five steps forward in any direction, puts the card down on the ground and stands by it. The child holding card number 10 walks to that card, then says 2 x 5, walks five steps forward in another direction, puts the card down on the ground and stands by it. The child holding card number 15 continues in the same way and so on to the child with card number 60.

If the space is large enough, the children can form a long line across the space. If the space isn't big enough for that to happen, the line can bend backward and forward and around itself to fit. Ask each child to change direction at least slightly, so that the line is not straight, making it easier to count up in 5s.

Extension/challenge
Play the game in the same way, but work in descending order, beginning with the player who says 12 x 5 = 60 instead of the one who says 1 x 5 = 5.

Learning objective
Recognising the standard units and intervals within the 5 times table.

Preparation
Number cards for multiples of 5 from 5–60.

Number of players
12 players at a time.

Shake it up

5 times table

Activity
Call out the first part of a multiplication sentence, for example: 6 x 5. Children take 6 piles of 5 marbles from the first tray, put them all into the box, put the lid on tightly and shake it.

Then they open the box, empty the marbles out onto the second tray and count them. When they have agreed on the correct multiple, they replace the marbles in piles of 5 on the first tray, counting and checking the piles as they do so.

Extension/challenge
Put a different number of marbles into the box and hand it to the children for each turn. They tip out and count the marbles and guess the number of 5s, then sort them into piles of 5 to check their answer.

Learning objective
Understanding the relationship between groups of 5 and their products.

Preparation
- Two trays
- Plastic box with a tightly fitting lid
- 60 marbles (or counters or stones). Make 12 piles of 5 items on the first tray before the game starts.

Number of players
Any number. Children may take turns to work out answers on their own or in small groups, or the whole group can work together on each sum.

How many minutes?

5 times table

Activity
Ask twelve children to stand in a circle at regularly spaced intervals, so that they resemble the numbers on a clock face. Stand in the centre of the circle with the ball and throws it to each child in turn, starting at 5 past and moving around at 5 minute intervals.

Each child catches the ball and throws it back to you. During this, the whole class chants the multiplication sentences together: 1 x 5 = 5, 2 x 5 = 10, 3 x 5 = 15 and so on to 12 x 5 = 60.

Extension/challenge
Ask the children to throw the ball quickly to each other around the circle, chanting as they throw and catch it: 5, 10, 15, 20, 25, 30, 35, 40, 45, 50, 55, 60! Draw their attention to the fact that time is counted in 5 minute intervals in this way.

Learning objective
Remembering and practising factors, multiples and sums in ascending order to create the 5 times table.

Preparation
Large ball (foam if playing inside).

Number of players
Ideally 12 players, or 24 players can create two circles and play simultaneously. If there are more children, they may stand in the centres of the circles and throw the balls, helped by an adult where necessary.

We go together

5 times table

Activity
Pick up a number from the pile and choose a 0 or a 5 card to put with it to form a multiple from the 5 times table. For example: 20.

Children discuss the factors and form themselves into a group of 5 and a group of the other number needed, eg 4 in the above example. They present the groups to you and say the sum aloud: 4 x 5 = 20.

Continue to choose cards to make multiples from the 5 times table; the remaining children rush to form themselves into the correct sized groups. Choose the multiples carefully, according to the number of players, to ensure that all children have enough opportunities for turns. For example, if there are 30 players:

You choose	To accommodate
25	10 players (5 x 5)
35	12 players (7 x 5)
15	8 players (3 x 5)

Learning objective
Linking factors and multiples within the 5 times table.

Preparation
- Six '0' (zero) cards and six '5' cards, placed in two separate piles, face up
- Cards with the following numbers on them: 0, 1, 1, 2, 2, 3, 3, 4, 4, 5, 5, 6, shuffled together and placed in a pile face down.

Number of players
Best played with 30 players.

This way every child would be involved; they could break apart for a second round, such as:

You choose	To accommodate
45	14 players (9 x 5)
05	6 players (1 x 5)
25	10 players (5 x 5)

The cards that make up the multiples should be laid out on the floor as they are used, as a reminder of the sums already solved. You must ensure that each of the 12 answers will be used once, by choosing the 0 or the 5 (whichever was not used before) when picking up a digit for the second time. For example: 10 and then 15 or 35 and then 30. Draw the children's attention to the fact that all of the answers end in 5 or 0.

Extension/challenge
Place a 0 card on the floor on one side of the space and a 5 card on the other. Call out factors in the 5 times table and children choose whether they think that the correct answer multiple ends with a 0 or a 5 for each one. They should walk quickly to the side they choose each time, not watching or copying each other. Announce the multiple to confirm the correct answer for each sum and praise those who went to the right side. (Don't award points or allow those who did not get the answer correct to lose confidence or stop trying. Move quickly onto the next number and assure the children that everyone will get some answers right and some wrong during practice.) This is a great activity for assessing which children still need more practice to develop confidence.

Inner circles

5 times table

Activity

Divide the children into two groups. Call out a sum from the 5 times table. For example: 5 x 5 = 25. Ask one group to take out all the straws that are not needed in the first hoop, to leave 5 there. Then ask the other group to take out all the straws that are not needed in the second hoop. For example: to leave 25 there. If the group is of mixed ability, those who are more able or confident should be in the second group.

Extension/challenge

With the children in two groups as in the activity above, call out only the multiple and ask players to place straws in their hoop to represent the factor that goes with 5 to create that multiple. For example: call out 55 and ask the first group to decide the correct number of straws to place in their hoop. Suggest that they think "How many 5s make 55?"

Remove the straws from the hoops, then call out only the number of 5s and ask players to place straws in their hoop to represent the multiple that will be formed when that factor is combined with 5. For example: call out 7 and ask the second group to decide the correct number of straws to place in their hoop. Ask, 'What is 7 x 5?'

Learning objective
Recognising how to record the different parts of the 5 times table.

Preparation
- 3 cards showing x, 5 and = . Lay out the three cards on the ground to form a multiplication sentence
- 2 hoops, one at each end of the sentence
- 80–100 drinking straws – place 15+ straws in the first hoop and 65+ straws in the second hoop.

Number of players
4+ players. As there will be only one chance to form each sum, they must all watch and pay attention throughout each turn in order to learn the whole 5 times table.

Fun Games and Activities for Teaching Times Tables

Pile it on

5 times table

Activity
Ask each group to work together to put their coins onto the squares of the number line in piles of one coin, two coins, three coins and so on up to twelve coins, discussing the multiple total that each number of 5p coins makes.

Then ask them to chant the sums aloud in order, as they count and check the piles of coins. For example: 1 x 5 = 5, 2 x 5 = 10, 3 x 5 = 15 and so on to 12 x 5 = 60.

Extension/challenge
Give the group multiples from the 5 times table in a random order. The children quickly decide how many 5s are in each multiple and present that number of coins in a pile, placing them on the correct squares of the number line.

Learning objective
Using knowledge of the 5 times table when counting coins and working with money.

Preparation
- Number line showing the multiples of the 5 times table (eg 5, 10, 15, etc) with squares large enough to hold piles of small 5p coins.
- 5p coins (78 coins per group) – either real coins (bags of coins are available from banks), play money or make rubbings of 5p coins stuck onto small circles of cardboard. (If using plastic coins or rubbings for the activity, provide a few real coins for the children to examine and handle first to ensure that they will recognise them in real situations.)

Number of players
The number of coins or rubbings that can be obtained or made will determine the number of players, but aim for 3–6 in a group and either 1 or 2 groups with an adult. If a large quantity of coins or rubbings is available, it is possible for more groups to play simultaneously with different adults.

© Debbie Chalmers and Brilliant Publications · Fun Games and Activities for Teaching Times Tables

This page may be photocopied for use by the purchasing institution only.

Towering heights

10 times table

Activity

Ask the whole group of children to work together to build a line of towers from blocks. The first will be only 1 block high, the second 2 blocks, the third 3 blocks and so on to 12 blocks. Discuss the relative sizes of the towers that increase in height by the same amount with each additional block, and count them several times 1–12.

Next, ask children to build towers that are made from 10 blocks, 20 blocks, 30 blocks and so on to 120 blocks. Unless a very large number of blocks are available, the smaller towers will need to be dismantled to build the taller ones, so children must be encouraged to remember them or to take photographs that they can display and refer to.

As towers are completed, count the numbers of blocks in them and record them in order. Children will be able to see and say the numbers: 10, 20, 30, 40, 50, 60, 70, 80, 90 100, 110, 120.

Learning objective
Understanding that counting on in tens in ascending order, to make larger and larger numbers, creates the 10 times table.

Preparation
A set of large blocks or smaller bricks, containing many identically sized and shaped pieces.

Number of players
Any number of players may take part in this game, but the group should be small enough to ensure that every child will have something to do.

Extension/challenge

Move this activity to tabletops and encourage children to work in smaller groups. Offer construction kits with many identical bricks or similar pieces and allow them to create their own sets of 12 small towers with 1–120 pieces.

Jump and turn

Activity

Ask children to form circles with 10 children in each. The children in the first circle hold hands and walk around, saying together as they turn: 1 x 10 = 10. The children in the second circle hold hands and both circles walk around: saying together as they turn: 2 x 10 = 20. The children in the third circle hold hands and the three circles walk around: saying together as they turn: 3 x 10 = 30.

One pair of children in the first circle break hands and the circle walks to become a line. The children in the line jump four times together, saying: 4 x 10 = 40. The children in the second circle repeat this but jump five times, saying: 5 x 10 = 50. The children in the third circle repeat this but jump six times, saying: 6 x 10 = 60.

Seven children in the first line remain standing while the other three sit down, then the three jump up and they all jump together as a line of ten. While they are doing this, they say: 7 x 10 = 70. Eight children in the second line remain standing while the other two sit down, then the two jump up and they all jump together as a line of ten. While they are doing this, they say: 8 x 10 = 80. Nine children in the third line remain standing while the other one sits down, then the one jumps up and they all jump together as a line of ten. While they are doing this, they say: 9 x 10 = 90.

The ten children in the first line sit down, while the other two lines stand facing each other. The lines jump ten times each and all of the children say together: 10 x 10 = 100. One child from the first line joins the second line, to make eleven children, and they stand facing the third line of ten children. The lines jump eleven times and ten times and all of the children say together: 11 x 10 = 110. Two children from the first line join the second line and stand facing the third line of ten children. The lines jump twelve times and ten times and all of the children say together: 12 x 10 = 120.

Extension/challenge

Children form three circles with 10 children in each. Ask the first circle to become as small as possible, by standing very close together. Ask the second circle to stand outside the first circle, holding hands to surround the children in the centre. Ask the third circle to stand outside the second circle, stretching to reach each other's hands and enclose the two smaller circles. Draw their attention to the fact that multiplying each number by 10 in turn makes the total bigger by another 10 each time and that creates the 10 times table. Ask them to imagine another 10 and another and another outside their circles, making the total bigger and bigger.

Learning objective
Understanding that counting on in tens in ascending order creates the 10 times table.

Preparation
No equipment needed.

Number of players
30, ideally, as they can form three circles. If there are fewer, adults may play too.

What's the catch?

10 times table

Activity
Ask children stand in one, two or three circles, each containing 6 to 12 children. Children bounce the ball to each other around the circle, counting how many bounces and catches are achieved before someone misses it. If they get to 12, they start again at 1.

The number reached when a catch is missed is the number the group must count to in the 10 times table and discover the answer to the multiplication sentence. The whole sentence or just the multiple may be said, eg the children could count together: 1 x 10 = 10, 2 x 10 = 20, 3 x 10 = 30, etc or they could just count: 10, 20, 30, etc.

Learning objective
Remembering and practising chanting the sums rhythmically in ascending order to create the 10 times table.

Preparation
1–3 large balls (foam if playing inside).

Number of players
Any number from 6 to 36.

The bouncing and catching begins again each time the answer is reached. Players should aim to include each multiple at least once, but it is not possible to fix the game to make this happen and sometimes that would take too long.

Extension/challenge
Play the game in the same way but, when a player misses the ball, they must say the correct multiple answer. For example 30 or 90 or 110. Each player must guess the multiple on their own as quickly as they can, without counting up from 10, and then bounce the ball to the next player for the game to continue.

This means that all players must concentrate all the time, to be sure of which number of catches have been achieved and the number of the missed catch, as well as working out the multiple answer.

Fun Games and Activities for Teaching Times Tables © Debbie Chalmers and Brilliant Publications

Count up and count down

Activity
Give out one 0 card each to 12 children and ask them to stand together in a group. Ask 21 other children to form groups of 6, 5, 4, 3, 2 and 1 and give them each the correct number card for their group.

The child representing number 1 walks to the group of children with the 0 cards and takes a card from one of them. Child number 1 says: 1 x 10 = 10 and lays the 2 number cards, 1 and 0, on the floor as 10. The two children in group number 2 walk to the group of 12 children and take a 0 card, then say: 2 x 10 = 20 and lay the 2 cards on the floor as 20. This continues to 60.

Learning objective
Visualising and practising counting on in tens in ascending and descending order to create the 10 times table.

Preparation
- Number cards 1-12
- 12 or 13 '0' (zero) cards.

Number of players
Ideally 36. If there are fewer children, adults may join in with the groups.

At this stage the children will need to form new groups for numbers 7 and 8 (15 children), then reform again for numbers 9 and 10 (19 children) and for number 11 and number 12. While children are not in play, they can help with the chanting of the sums. When all of the numbers are laid out on the floor together, ask the children what they notice.

Extension/challenge
Ask children to pick up the cards one by one, in descending order, and chant the names of the multiples, counting backwards. For example: 120, 110, 100, 90, 80, 70, 60, 50, 40, 30, 20, 10.

If desired, make an extra 0 card to bring out at the end and allow the children to count down to zero and blast off around the room as rockets!

Meet in the middle

Activity
Give the set of 1–12 number cards to one child and the set of 0 cards to another child. Give 10 drinking straws each to 12 other children.

Call out a multiple from the 10 times table. The correct number of players with straws come together in the centre to create the multiple answer to that multiplication sentence, while the player with the number cards brings the factor number called and the other player brings one of the 0 cards. They all stand together to create the number.

For example: you call 4 x 10. Four children bring their 10 straws each and the other two children bring the number 4 card and a 0 card. Together, they decide that the answer is 40 and state this for the group.

The players with straws may stay together while others take their turn to form the next number. The players with cards will leave the first ones with the small group and move to the next group to supply the right cards for them. When there are not enough players left to create the multiple for the multiplication sentence called, some may leave the groups they have already been in and move to others to make up the correct groups.

Learning objective
Visualising and understanding the increasing quantities involved when creating the multiples of the 10 times table.

Preparation
- Number cards 1–12
- 12 '0' (zero) cards
- 120 drinking straws.

Number of players
14. Adults may join in with children to make up the right number within the group. If this happens, it is ideal if an adult takes the 0 cards, as supplying one of these to each group does not demand particular skills and teaches less than the other tasks, although being involved in the game will allow all children to learn the sums that they help to create.

Extension/challenge
Play the game with 24 children, as two teams of 12, and allow them to make up numbers for each other to guess. Give 10 straws to each player in the first team and the 1–12 and 0 cards to the second team. Ask the players in the first team to come together to hold straws while the second team chooses the right cards to create the multiple. After a few turns, allow them to swap places, so that the players in the second team use the straws and the first team uses the cards. You will only need to supervise and offer advice.

Magic rings

10 times table

Activity
Tell the children that the hoops are magic rings that can multiply numbers by 10. Call out a number from 1-12 and point to a child. That child should look for the correct hoop, jump into it, pick up the card and chant the sum. For example: Call 5 and point to a child. The child looks for the hoop with the 50 card in it, then jumps into the hoop, picks up the card and chants aloud 5 x 10 = 50.

When each child has had a turn, there should be one child standing in each hoop, holding a card. Play the game more than once, calling multiples and pointing to players in different orders, to ensure that they all have the opportunity to choose from a large selection of numbers and work out different sums.

Extension/challenge
Ask the children if they have noticed that 10 times a number is formed by placing a zero in the units column beside that number, which now moves to the tens column. Suggest that they each take a turn to jump in and out of the hoops in the order of the 10 times table, chanting the multiples as they jump: 10, 20, 30 … 120.

Learning objective
Recognising the multiples of the 10 times table and the factors they are linked to.

Preparation
- 10 hoops, laid out in a random pattern on the ground or floor
- 12 number cards showing the multiples in 10 times table: 10–120. Place one number card inside each hoop, face up, in a random order.

Number of players
Ideally 12. A smaller group can play if some or all players take two turns each.

Throw it aside

Activity
Children form two groups of 12 which stand on either side of the space. Number the children in each group 1–12. The two children representing number 1 (one from each side) take one stride towards each other, saying: 1 x 10. The number 2s take two slightly smaller strides towards each other, saying: 2 x 10. This continues with each number until the number 12s are taking tiny steps and ending up very close to each other to say 12 x 10.

Throw the balls to the number 1s and, as they catch them, they say 10. They then throw the balls to the number 2s in the opposing group who say 20 as they catch the ball. This continues until the number 12s catch the balls and say 120. The two balls will be flying back and forth across the space between the correct answers, making the game very stimulating.

Learning objective
Linking factors and multiples within the 10 times table.

Preparation
Provide two soft (foam) balls that are large enough for the children to throw and catch easily and heavy enough to travel some distance.

Number of players
Ideally 24. If there are fewer children, adults may join in. It is best played in a large open space.

Extension/challenge
Suggest to the children that they bounce the ball the correct number of times after catching it, counting aloud in multiples, before throwing it across to the next player. For example: when the number 6s catch the balls, they bounce them 6 times, counting 10, 20, 30, 40, 50, 60 and then throw them over to the number 7s in the other group.

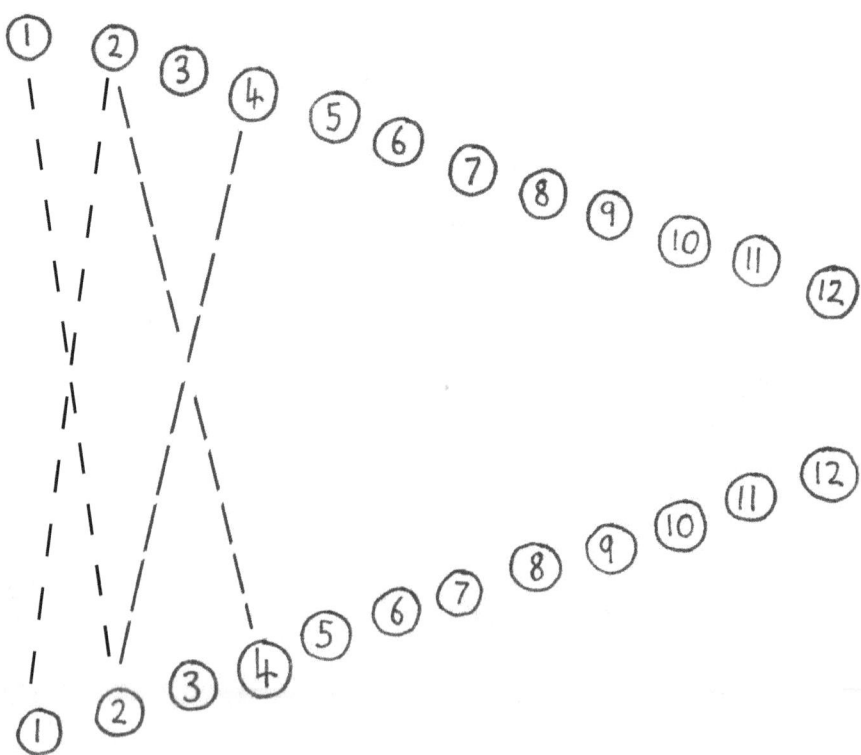

Fun Games and Activities for Teaching Times Tables   © Debbie Chalmers and Brilliant Publications

Catch this!

Activity
Ask children to stand in a circle and give them each a number 1–12. Stand in the centre of the circle with the pile of beanbags. Call out a multiple from the 10 times table and the children must work out the answer. The child who represents the correct factor must shout it out and you then throw a beanbag to them. They should catch it and sit down in their place. For example: you call 120 = 10 x ? The child who is number 12 shouts 12 and catches the beanbag that you have thrown to them and then sits down with it. If a child does not know the answer and doesn't respond when appropriate, other players may call out to encourage them to react in time.

When each child has had a turn, they will all be sitting down, holding beanbags. Ask the children to stand up, throw the beanbags back to the centre and shuffle into different places before playing again. Play the game more than once, calling out sums in different orders, to ensure that they all have the opportunity to represent different numbers.

Learning objective
Linking and practising the sums, factors and multiples of the 10 times table.

Preparation
- 12 beanbags
- Large outdoor or indoor playing space.

Number of players
Ideally 12. If the group is smaller, adults may join in. If enough beanbags are available, more groups can play simultaneously with more adults.

Extension/challenge
Ensure that every child knows which number they are representing, by chanting them around the circle once or twice before the game starts. Then play the game without any shouting out, so that children must hear the sum, think of the answer silently and react quickly to be ready to catch the beanbag when it is thrown to them. Wait for three seconds after calling out the sum and then throw the beanbag to the correct child, whether they are ready or not. If they are surprised or miss the catch, this will just add to the fun for the group.

Roll in order

10 times table

Activity
Set the skittles labelled 1–12 up in a random pattern. Ask each player in turn to roll the ten balls one by one and to try to knock down the skittles in ascending numerical order. If they miss one, they should leave it standing and continue to the next number with the next ball. After each turn, tell the player their score and then set up the skittles again.

When every player has rolled the balls, set up the skittles labelled 10–120 instead and repeat the game. Players must try to knock down these skittles in ascending numerical order, understanding that they are the multiples of the 10 times table.

Extension/challenge
Allow children to form teams of 2–4 players and roll the balls together, offering each other advice on how to knock down all the skittles in order. They may think of some interesting strategies and will need to look and plan carefully, which will also help them to absorb the correct order of the multiples.

Learning objective
Recognising the factors and multiples of the 10 times table in ascending order.

Preparation
- 12 skittles labelled 1–12
- 12 skittles labelled 10–120
- 10 suitable balls that can be rolled to knock skittles down.

Number of players
Any number, but the group should not be too large, to avoid children having to wait too long for their turns.

Fun Games and Activities for Teaching Times Tables

Going round in circles

Activity
Work with the group of children to create circles of money. Place one 10p coin in the centre of the table or floor space. Measure and cut a piece of string to encircle the coin and lay it down around it. Place two coins outside the string and measure and cut another piece to encircle them. Continue in this way with three coins, then four and so on up to twelve, creating strings that form ever increasing circles with coins between them. (The finished design will resemble a target).

Ask the children to take turns to pick up the coins, carefully without disturbing the strings, counting in 10s to find the multiples of the 10 times table. They should begin with the centre coin and work towards the outer ring. For example: pick up the centre coin and say 10, pick up the 2 coins in the next inner ring and say 20, and so on until picking up the 12 coins in the outer ring and saying 120.

Learning objective
Using knowledge of the 10 times table when counting coins and working with money.

Preparation
- A ball of string
- A few pairs of scissors
- 10p coins (78 coins per group) – either real coins (bags of coins are available from banks), play money or make rubbings of 10p coins stuck onto small circles of cardboard. (If using plastic coins or rubbings for the activity, provide a few real coins for the children to examine and handle first to ensure that they will recognise them in real situations.)

Number of players
1 or 2 groups of 2–6 players. If enough resources are available, more groups can play simultaneously with more adults.

Extension/challenge
Ask children to start with the outer ring and count backwards to the centre: 120, 110, 100 …10. When children are familiar with this game, ask them to pick up coins from particular rings, not in ascending or descending order. Either tell them the multiple and let them say the number that is the other factor with 10, or tell them a number of coins and let them find that ring and say the multiple.

© Debbie Chalmers and Brilliant Publications
This page may be photocopied for use by the purchasing institution only.

Fun Games and Activities for Teaching Times Tables

Build it up

2, 5 & 10 times tables

Activity
Divide the children into three groups.

Ask the first group of children to make towers of 2s by putting 2 blocks or bricks together as many times as possible, using all of the 20 pieces.

Ask the second group of children to make towers of 5s by putting 5 blocks or bricks together as many times as possible, using all of the 20 pieces.

Ask the third group of children to make towers of 10s by putting 10 blocks or bricks together as many times as possible, using all of the 20 pieces.

Talk with the whole group about the differences in the sizes and numbers of towers that the different groups have made from the same number of pieces.

Learning objective
Creating groups of two, five and ten.

Preparation
60 identically sized and shaped blocks or small bricks (20 for each group).

Number of players
6+ players working in 3 groups.

Extension/challenge
Ask the children to compare the towers that the different groups have made and to describe them to each other, exploring the mathematical relationships between their sizes and numbers of pieces.

For example:
- Two towers made from 5 pieces can be put together to make one wall of 10 pieces. There are ten smaller towers each made from 2 pieces, but two larger towers each made from 10 pieces.

- Each group has used 20 pieces so, if the ten small towers, the four middle-sized towers and the two large towers are pushed together, they will each create walls that are the same size.

Knock them down

2, 5 & 10 times tables

Activity
Ask the children to divide the skittles into 5 groups of 2 and set each pair up in a separate place. They roll the 5 balls to try to knock the pairs down one by one.

Children then divide the skittles into 2 groups of 5 and roll 2 balls to try to knock the groups down one after the other.

The skittles are then set up together as a group of 10 and children roll a ball to knock down as many as they can.

Extension/challenge
Allow the children to play in teams of 5, rolling one ball each and aiming to knock down every skittle in each of the three formations. They may advise each other on where to roll and how to knock down the greatest number at a time.

Learning objective
Dividing quantities and visualising groups of 2, 5 and 10.

Preparation
- Set of 10 skittles
- 5 balls of suitable size and weight that can be rolled to knock the skittles down.

Number of players
Any number.

© Debbie Chalmers and Brilliant Publications — Fun Games and Activities for Teaching Times Tables
This page may be photocopied for use by the purchasing institution only.

Change the shape

2, 5 & 10 times tables

Activity

Children stand beside each other and hold hands in a line across the room, making two lines of 10 players. The two lines walk (or skip) towards each other. The children in the first line hold their arms up high to form arches and the children in the second line drop hands and walk through the arches. Both lines turn around and hold hands again once they are through. They then repeat the movement with the second line forming the arches and the first line walking through them.

Learning objective
Recognising the differences between quantities of 2, 5 and 10.

Preparation
No equipment needed.

Number of players
Ideally 20 players, divided into 2 groups of 10.

Each of the two lines now splits in half and forms two lines of 5 players. The four lines then move towards the centre of the space, so that the 4 children who are each at one end of their line come together and put their hands together in a star formation. The lines walk around, turning like the spokes of a wheel. They then drop hands and the lines turn around, so that the 4 children who are each at the other end of their line can put their hands together in a star formation and the lines can walk around again, like the spokes of a wheel turning in the other direction.

Children leave their lines of 5 and form pairs instead, each group of 2 children holding hands with each other. The pairs line up behind each other and walk (or trot like horses) to promenade around the space in any directions, following the leading pair.

Discuss with the children the different ways of dividing a group and forming smaller equal groups. They may begin to recognise the relationship between the three numbers 2, 5 and 10.

Extension/challenge

Once the children are familiar and confident with each of the three formations, ask them to play the game by creating the differently sized groups and making the movements as they hear you call out the numbers, changing quickly when each new number is called, in random order.
For example: 10s, 5s, 2s, 5s, 2s, 10s, 5s, 10s, 2s.

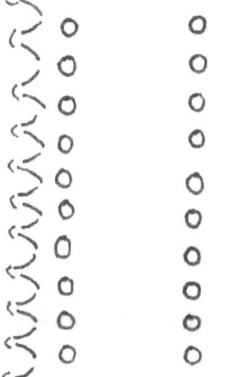 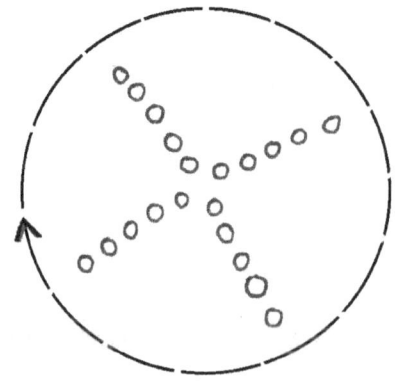 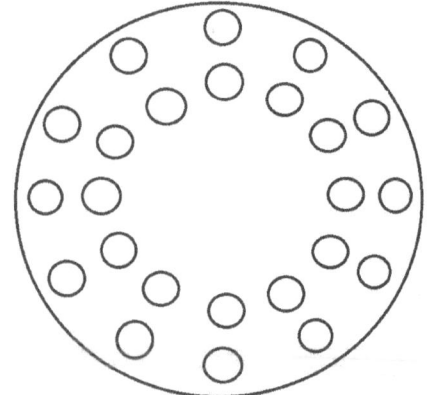

Pass it on

2, 5 & 10 times tables

Activity
Divide the children into groups of 2 and give one ball to each pair. Children stand in spaces and throw and catch the balls with their partners.

Next, children form groups of 5 and stand in lines. Give one ball to each line. Children throw and catch the balls, passing them up and down the lines.

Finally, children form groups of 10 and stand in circles. Give one ball to each group. Children throw and catch balls around the circles.

Learning objective
Understanding the differences between groups of 2, 5 and 10.

Preparation
Soft (foam) balls that are large enough for the children to throw and catch easily (5 balls for each 10 children taking part).

Number of players
10, 20 or 30 players, divided into groups of 10.

Extension/challenge
Place the balls in a large container in the centre of the space and ask the children to stand around it. Call out the numbers 2, 5 and 10 randomly and children must form the correct groups, take the right number of balls and begin the throwing and catching exercises by themselves. Call out a new number at regular intervals, after all children have participated in a few throws and catches, and the children must change their groups appropriately each time.

We all stand together

2, 5 & 10 times tables

Activity
Divide the children into groups of 10.

Ask the first group of children to make groups of 2 and stand together in pairs. Ask the second group of children to make groups of 5 and stand together in two halves. Ask the third group of children to stand together as one group of 10.

The children forming the group of 10 stand in a line, one behind each number card. The children forming the groups of 5 stand behind them, proving that 2 groups of 5 standing beside each other equal the same amount as 1 group of 10. Children forming the groups of 2 then stand behind them, proving that 5 groups of 2 standing beside each other equal the same amount as the 1 group of 10 and the 2 groups of five.

Explain to the children that these relationships are a part of the times tables.
For example: 1 x 10 = 10 and 2 x 5 = 10 and 5 x 2 = 10.

Extension/challenge
Describe the line of 10 as a whole one and explain that the other groups are fractions of the whole. Discuss halves and fifths. Suggest that one group of 5 steps away and ask the children to say what fraction is now missing from that line (half). Suggest that one group of 2 steps away and ask the children to say what fraction is now missing from that line (one fifth). Also try this with two, three and four fifths.

Learning objective
Recognising the relationships between the numbers 2, 5 and 10.

Preparation
Number cards 1-10 (2–3 sets). Lay out the number cards on the floor in a line, reading 1–10 from left to right.

Number of players
10, 20 or 30 players, divided into groups of 10.

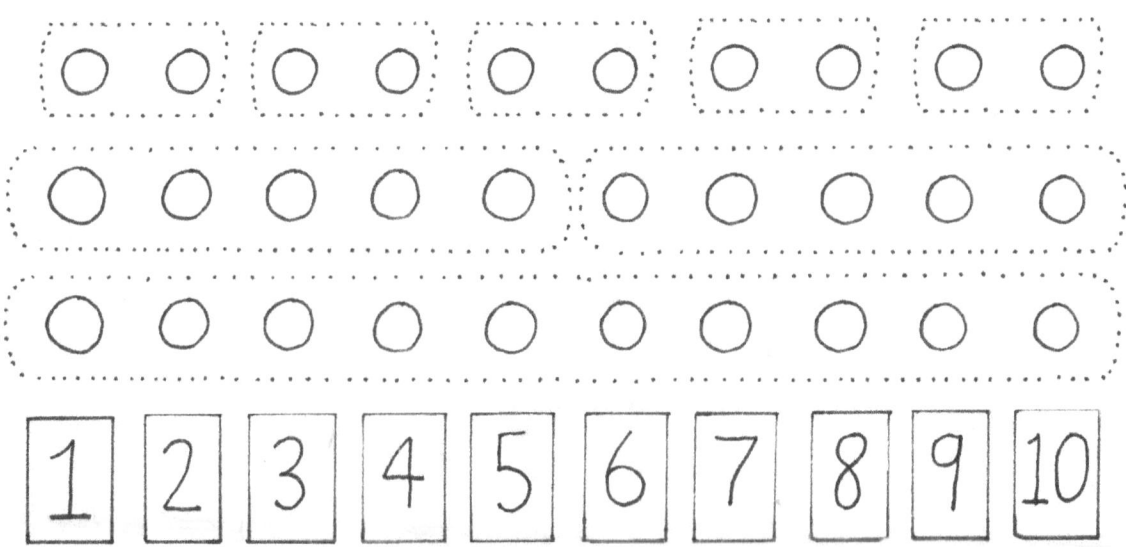

In the ring

2, 5 & 10 times tables

Activity
Ask the children to throw 2 beanbags into each hoop and count how many hoops are filled when all 20 beanbags have been used.

Children then take the beanbags back and throw 5 into each hoop until they have all been used, then count how many hoops are filled.

Finally, they gather the beanbags again and throw 10 into each hoop, to find out how many hoops are filled in this way.

Discuss with the children that 10 x 2 and 4 x 5 and 2 x 10 may all be made using 20 beanbags.

Learning objective
Discovering that the same quantities can be divided into groups of 2, 5 or 10.

Preparation
- 10 hoops laid out in a random pattern on the floor or ground
- 20 beanbags.

Number of players
Any number.

Extension/challenge
Take three beanbags away, leaving seventeen for play. Challenge the children to throw 2 into one hoop, 5 into another hoop and 10 into a third hoop. Ask if they can count each set of beanbags once, then twice, then three times to begin to chant each of the tables. For example: 2, 4, 6 or 5, 10, 15 or 10, 20, 30.

Break it down

2, 5 & 10 times tables

Activity
Divide the children into groups of between two and six players and give each group a model, wall or tower.

Ask the groups to break their models down into smaller models that are each made up of 10 bricks and to find out how many models they have. Then ask them to break these models down into smaller ones that each have 5 bricks and say how many they have. Finally, ask them to use all the bricks to make small models with 2 bricks each and decide how many models they can make in this way.

Count the bricks together and explore the relationships:
30 = 3 x 10 or 6 x 5 or 15 x 2.

Extension/challenge
Ask pairs or small groups of children to explore the numbers 20 and 10 independently, in the same way as in the game above.

Learning objective
Understanding how multiples are created within the 2, 5 and 10 times tables.

Preparation
Build a number of completed models, walls or towers, each with 30 pieces, using small bricks or construction kits. The number of models needed will depend upon the number of players.

Number of players
Any number. If the group is small, they will work in pairs; if it is larger, they will work in larger groups.

Fun Games and Activities for Teaching Times Tables

Gone shopping

2, 5 & 10 times tables

Activity
Invite the children to take turns to buy items from each shop. Give them a pile of 2p coins and ask them to choose an item from the first shop, then find the right number of coins to pay for it. Repeat this with the 5p coins for the second shop and the 10p coins for the third shop.

Extension/challenge
Give the children piles of each of the three different coins at the same time and allow them to choose items from any of the shops and find the right coins to pay for them.

Learning objective
Recognising the multiples that appear within the 2, 5 and 10 times tables.

Preparation
- 3 tables or mats to be shops. On each, place a collection of items, such as books, balls or empty boxes, marked with prices. The set of prices in the first shop should be the multiples of the 2 times table, ranging from 2p to 24p. The set of prices in the second shop should be the multiples of the 5 times table, ranging from 5p to 60p. The set of prices in the third shop should be the multiples of the 10 times table, ranging from 10p to 120p. (Do not introduce £ as well as p, as it would confuse this activity.)
- 2p, 5p and 10p coins (as many as possible – these may be real coins, play money or rubbings of coins cut out and stuck onto card circles).

Number of players
Any number. If there are not enough coins for each player to have their own, they can play in small groups or take turns to use the same coins.

© Debbie Chalmers and Brilliant Publications
This page may be photocopied for use by the purchasing institution only.

Fun Games and Activities for Teaching Times Tables

Stand over there

2, 5 & 10 times tables

Activity
Tell each child in turn a multiple number and ask them to go to stand by the correct factor card. If they are not sure, they may use the bricks to count and check.

None of the multiples appear in just two of these three tables, but the numbers 10 and 20 appear in all three. You may wish to avoid using these, or allow the children to decide where they would like to stand. Point out to the group that these numbers link the three tables, if they do not realise it for themselves.

Extension/challenge
Play with a small number of children and invite them to respond as a group to every multiple that you call out, running between the three cards. If they hear one of the numbers that links the three tables, they should sit down on the floor or ground instead.

Learning objective
Linking multiples to the 2, 5 and 10 times tables.

Preparation
- Number cards 2, 5 and 10 stuck on walls in three different places (or ask three adults to hold them)
- 2 bricks, 5 bricks and 10 bricks in another place, to be used as visual aids.

Number of players
Any number. If the group is small, they may take more than one turn each and move from one card to another.

Fun Games and Activities for Teaching Times Tables   © Debbie Chalmers and Brilliant Publications
This page may be photocopied for use by the purchasing institution only.

Keeping time

2, 5 & 10 times tables

Activity
Ask the children to take turns to jump around the outside of the clock, landing at 10 minute intervals, chanting 10, 20, 30 …

Repeat with 5 minute intervals, chanting 5, 10, 15, 20 … Then repeat with 2 minute intervals, chanting 2, 4, 6, 8 …

Explain that 5 minute intervals are usually used to help people to tell the time, as we say: 5 past 4, 20 past 6, 25 to 10 or 10 to 12, or we say: 4.05, 6.20, 9.35 or 11.50. But we might often measure time in 10 minute intervals because it is convenient for some periods, such as appointments or breaks. We might sometimes need to be more accurate and so might say 2 minutes past 7 or 8 minutes to 9.

Extension/challenge
Talk about the clock being divided into four quarters and how time is described as: a quarter past 5, half past 6, a quarter to 9 and 10 o'clock. We do not usually say: 15 past 5, 30 past 6, 45 past 8, 15 to 9 or 0 past 10, but we do write and say: 5:15, 6:30, 8:45 and 10:00.

Ask children to identify and find these quarters and then to consider how, since they are each 15 minutes long, 3 x 5 minute intervals fit exactly into each one, but 10 and 2 minute intervals do not.

Learning objective
Understanding how intervals of time can be measured using the 2, 5 and 10 times tables.

Preparation
Large clock face, either chalked on the playground outside or drawn out on a large sheet of paper and taped to the floor. Mark out all the minutes 0-60 with numbers and lines at 5 minute intervals.

Number of players
Any number. Children can take turns to jump one after the other, but, if the group is large, they may jump in pairs or threes.

© Debbie Chalmers and Brilliant Publications — Fun Games and Activities for Teaching Times Tables

Toss it over

Learning in sequence

Activity
Ask the children to sit separately in spaces and tell them which times table is being studied. Chant the sums, one by one, in ascending order. For example:

  1 x 12 = ?
  2 x 12 = ?
  3 x 12 = ?

Point to a different child as you ask each question. The child stands up, goes to the pile, takes the right number of beanbags (1, 2 or 3) and throws them into the buckets that show the correct numerals, in the correct order, to form the answer. For example: ask the question 9 x 12 = ? and the child takes 3 beanbags and throws them into the buckets labelled 1, 0 and 8.

Extension/challenge
Ask children to find the answers for the 2 times table and then the 3 times table, throwing the beanbags into the buckets as before, and to discover how many multiples and how many beanbags they have to count before they have thrown a beanbag into each bucket at least once. (The answer is 21 multiples and 35 beanbags, because they will need to go through the 2 times table up to 12x and then up to 9 x 3 = 27 in order to use the last bucket, which will be the number 7.)

Learning objective
Visualising the multiples within one or more complete times tables in correct ascending order sequence.

Preparation
- 10 plastic buckets, boxes or crates labelled with numerals 0–9, set out in a line across the space, in correct numerical order.
- Beanbags. The number needed will depend upon the table(s) chosen and should be worked out in advance. (The 2 times table requires 20 beanbags, but the 12 times table requires 28.) Place the beanbags in a pile on the floor.

Number of players
6+ players. 12 players will take one turn each; 6 players will take two turns each per times table (but more than one table may be chosen). If the number of players does not divide into 12, each player must take at least one turn, but some may be chosen at random to take more than one. If there are 24 or more players, work on two or more different tables, one after the other.

Fun Games and Activities for Teaching Times Tables © Debbie Chalmers and Brilliant Publications
This page may be photocopied for use by the purchasing institution only.

Make a list

Learning in sequence

Activity
For this activity pupils work together as one group, discussing ideas and answers before placing the number cards.

Explain to the children that their task is to choose all the correct unit digits (and tens where needed) and put them in place, creating the times table multiples beneath each other in ascending sequence. Put the first multiple in place, with the digits in the correct column.

Scatter all the other cards over the ground or floor, face up, to be chosen in order by the children. Include a few red herrings and extra digits to make the end as difficult as the beginning and to prevent the players from working out the answers by elimination rather than knowledge of the table.

Extension/challenge
Scatter all the cards in two or three separate piles and tell the children which is the units pile, which is the tens and which is the hundreds (if needed). Only put the first multiple in place. For example: 0 6 or 1 2.

Learning objective
Visualising and ordering the multiples within one complete times table in ascending order sequence.

Preparation
- 2–3 cards showing U, T (and H if needed) to indicate places for units, tens (and hundreds), laid out on ground to indicate where digits will be placed
- Number cards showing the single digits needed to create all the multiples of the chosen times table. For example, for the 7x table: 7, 1, 4, 2, 1, 2, 8, 3, 5, etc, as well as some extra digits.

Number of players
Any number of players may take part.

Players must work together to place all the unit, tens (and hundreds) cards, to create all the multiples of the chosen times table in ascending sequence.

Turn and turn again

Learning in sequence

Activity
Ask 12 children to stand in a circle, holding up the cards 1–12, in a random order. Choose one child to stand in the centre of the circle, holding up the number card to indicate the times table chosen. For example: 3. Up to ten other children should hold one or two of the cards 0–9 each and stand in random spaces outside the circle.

The children in the circle walk around the child in the centre, chanting 1 x (3) is …. The child holding the numeral 1 and the child from the centre then break out of the circle together and walk to fetch the two children with the correct numeral cards for the answer from the children outside the circle. (This game requires numbers to be recorded as 01–09, then 10–12.) The other players call out to help. The children hold up the set of cards in order and all chant together. For example: 1 x 3 is 3.

All children then return to their places, still holding their cards. The child holding the numeral 2 and the child from the centre then break out together and repeat for the next question. The game continues in this way until 12 x 3 is ? is reached and solved. In the higher tables, three numeral cards will have to be fetched for the 3-digit numbers.

Extension/challenge
Have available a set of number cards 2–12 and give the child in the centre of the circle a numeral at random, then ask the children to play the game working on that times table. When they reach the end, ask them all to swap places and a different child to be in the centre of the circle and offer another random numeral, for the game to begin again.

Learning objective
Creating the multiples within one or more complete times tables in correct ascending order sequence.

Preparation
- Set of number cards 1–12
- 2 sets of number cards 0–9
- Number cards to represent the times table chosen.

Number of players
Ideally between 18 and 23 players

Fun Games and Activities for Teaching Times Tables

Count and catch

Learning in sequence

Activity
Choose a times table and ensure that you have a correct number of players to form two, three or four groups of that number. For example: If choosing the 6 times table, play with 12, 18 or 24 children (or adults if necessary). Ask each group to stand in a circle and number them: 1, 2, 3… .

Ask Group 1 to throw the ball to each other around the circle and count aloud as they catch it: 1, 2, 3, 4, 5, 6. When each player has caught the ball once, the group chants the first sum of the table: 1 x 6 = 6. The last player, who is holding the ball, then throws it to Group 2 and they repeat the counting around in the circle: 7, 8, 9, 10, 11, 12 and the chanting: 2 x 6 = 12.

The game continues in this way until the last multiple is reached: 12 x 6 = 72. If there are only two groups, Group 2 throws the ball back to Group 1. If there are three groups, Group 2 throws the ball to Group 3 and then Group 3 throws it back to Group 1. If there are four groups, Group 3 throws the ball to Group 4 and Group 4 throws it back to Group 1. The game is unaffected by the number of groups participating, as all players can watch, count and chant together, whoever is throwing the ball.

Extension/challenge
Bounce the ball from player to player, instead of throwing and catching it. Ask a player from the next group to be ready to run into the circle to catch the ball after the last bounce each time and take it back to their circle to continue the bouncing. This should be as continuous and seamless as possible and will allow the game to be played at a faster pace.

Learning objective
Working out and chanting the sums rhythmically in ascending order to create one complete times table.

Preparation
Ball that is large enough for the children to throw and catch easily (foam ball indoors, football outdoors).

Number of players
Any number of players may take part.

Into the middle

Learning in sequence

Activity
Ask the children to stand in a circle and tell them which table they are working on. For example: the 8 times table. Give them each the correct number of straws. Point to one child, who goes into the middle of the circle and lays down the 8 straws, chanting the sum aloud: 1 x 8 = 8.

This player then points to another child and rejoins the circle. The second player goes into the middle of the circle and lays the (8) straws down beside the first group of straws, chanting the next sum aloud: 2 x 8 = 16. The game continues in this way until the twelfth child has laid down the last pile of (8) straws to make the final total. For example: 12 x 8 = 96.

If a player does not know the answer while taking a turn, others may supply it. If no child knows the answer, the straws must be counted aloud by the group in unison, pile by pile, until the correct multiple is discovered.

Learning objective
Creating the sums and the multiples within one complete times table in ascending order sequence.

Preparation
Count out and prepare the number of drinking straws needed, depending upon the times table chosen. For example: the 2 times table will need 24, but the 12 times table will need 144.

Number of players
Ideally, 12 players. If there are more, they may work in pairs. If there are fewer, they may hold a pile of straws in each hand and take two turns each.

Extension/challenge
Ask children to (politely) take straws from each other to create their totals, by asking the correct number of players in turn: Please give me your 8 (or whichever number they are working on).

They should count and decide when they are right and chant the sum aloud for the group. For example: 4 x 8 = 32. They must then give the correct number of straws back to each player at the end of the turn, so that they can always be multiplied correctly.

How far can you go?

Learning in sequence

Activity
Ask 12 children to hold cards 1–12 and another 11 children to hold cards 2–12. If you have more than 23 children in the group, the other players will be in charge of handing out the correct cards 0–9 from the piles; if not, children may take them for themselves.

The children holding cards 1–12 and those holding cards 2–12 form pairs and walk together to find the answers for their questions, then take the correct 0–9 numerals to a designated space and lay them out in order. Beginning with the 2 times table, guide them to work their way through the tables, including each one that they have studied so far and stopping when they reach the first one that they have still to learn.
For example:
  1 x 2, 2 x 2, 3 x 2, 4 x 2 …
  1 x 3, 2 x 3, 3 x 3, 4 x 3 …
  1 x 4, 2 x 4, 3 x 4, 4 x 4 …

> **Learning objective**
> Creating sums within several complete times tables in correct ascending order sequence.
>
> **Preparation**
> ☐ Set of number cards 1–12
> ☐ Set of number cards 2–12
> ☐ 13 sets of cards 0–9, sorted and placed in 10 separate piles on the floor.
>
> **Number of players**
> 23+ players needed. Adults may join in as necessary.

Use your knowledge of the children's abilities, personalities and preferences when allocating parts, as obviously some numbers will have more work to do than others until all tables to 12x have been learned. If a large space is available, it is ideal to leave all of the answers displayed in separate lines until the end of the game. However, if space is more limited, each line can be gathered into a pile once you and the pupils have looked at it together.

Extension/challenge
Call out tables in a random order and ask children to work on them as they hear them. Children will still work out the answers to each table from 1x to 12x in correct ascending sequence, but instead of beginning with 2 and working methodically up to 12, they will need to call each table to mind separately and adjust their thinking more quickly.

Take your partners

> **Learning in sequence**

Activity
Give 23 children one card each from the 1–12 and 2–12 set and ask them to form the question lines at one side of the room or play area. The other children hold the cards 0–9 (two of each; one in each hand) and stand in the answer line at the other side.

The whole group chants the chosen table in sequence and, as each question and answer is given, the two children from the question line and the one, two or three children from the answer line walk to the centre and join together in the correct order. For example:

      6   7   42   or
     11   12   132

Learning objective
Creating sums within several complete times tables in correct ascending order sequence.

Preparation
- Set of number cards 1–12
- Set of number cards 2–12
- 2 sets of number cards 0–9.

Number of players
33 players. Adults and children can play together.

All children walk back to their places as the next chant starts and the next children walk to the centre to create the next phrase. This means that the child representing the numeral of the chosen table will walk each time, so it is ideal to repeat the game with a few different tables at each session, allowing several different children the opportunity to take on this role.

Extension/challenge
Ask the children to listen carefully, maintain concentration and be ready to react throughout the game, so that they can chant a whole table quickly but with a steady and rhythmic pace and move correctly for each phrase, without having to stop, wait or slow down.

In a mixed ability group, children can be encouraged to help each other, offering prompts and advance warnings where necessary to ensure that the group can work and cooperate efficiently as a team.

Very able children may particularly enjoy this game when they are working separately as an extension group and may achieve an impressive level of speed and competence.

Ever increasing circles

Learning in sequence

Activity
Ask the children to stand in a line and take turns to jump into each hoop in turn, counting: 1, 2, 3, 4 … . Call out numbers for when they should stop: 1 for the first child, 2 for the second child and so on up to 12. When they stop, they need to say the answer to that times table sum. For example: 4 x 9 = 36. Then they can turn over the card to see if they are correct. At the end of their turn, they step out of the hoop and stand beside it, to allow the next player to jump into it as they pass.

It will be necessary to give out the places in the line sensitively, so that the less confident children take the numbers 1, 2, 3, 5, 10 and 11, while the more able ones take 4, 6, 7, 8, 9 and 12. If the children's interest lasts, ask them all to return to the line after each player has had one turn and repeat the jumps. This time call out the numbers in a random order and ask the players to guess the sums as before, then chant the multiples in sequence as a whole group at the end.

Learning objective
Linking factors and multiples within one or more times tables in ascending order sequence.

Preparation
- 12 number cards with the correct multiples for the times table chosen, for example: 9, 18, 27 …108
- 12 hoops laid out in a line with number cards placed in them, face down, in correct ascending order.

Number of players
Ideally, 12 players will take part in this game. If 24 players and hoops are available, try preparing more cards and having two different times tables beside each other, so that two sequences can be worked on at once.

Extension/challenge
Lay out two or more sets of 12 hoops in lines, with number cards, to represent different times tables. Ask children to take turns to make a number of jumps, but allow them to choose in which directions to make the jumps, moving from line to line and back again whenever they wish.

When they stop, they must identify which sum in which times table they have reached and say it aloud. For example: if the 7 and 11 times tables are laid out, a player making 10 jumps may move through 1, 2 and 3 x 7, then over to 3 and 4 x 11, then back to 4, 5, 6, and 7 x 7 and then over to 8 x 11. By remembering which times tables they are now representing and counting up from the beginning of the line of hoops, they will know that they must say: 8 x 11 = 88.

Stand in line

Learning in sequence

Activity
Ask 12 children to stand in a spaced line, holding up the cards 1–12 in correct ascending order. Ask other children to form a group of between 2 and 12, depending on the times table chosen. Those left will have the task of finding the right numbers for the answers. Using the set of number cards, lay out the answers to the chosen times table on the floor. For example: if you have a group of 20 children practising the 5 times table, choose 12 to stand in a line and 5 to form the group, asking 3 to be the finders who stand with the answers. Take the correct number of cards and lay them out on the floor as one single card and eleven pairs of cards to form 5, 10, 15 … 55, 60.

The group walks to number 1 in the line and they chant together: 1 x (5) is ? They should go on to immediately say the answer, correcting each other if necessary. The group walks to the numbers on the floor and the finders pick up the card(s) to give them the right answer, helped by everybody else calling out to them. The group walks back to number 1 and gives that child the answer to hold. Then the group walks to number 2 in the line. The process is repeated with each number in turn, up to 12. When all the answers have been found, the whole group chants them together, in correct ascending order sequence, as the children in the line hold them up in turn. Aim for a rhythmic and enthusiastic shout. For example: 5, 10, 15, 20, 25, 30, 35, 40, 45, 50, 55, 60.

Learning objective
Linking the factors and multiples within one complete times table in correct ascending order sequence.

Preparation
- Set of number cards 1–12
- Sets of 20–28 number cards to form the correct multiples for the times table chosen, for example: 5, 1, 0, 1, 5 … 6, 0. Add some red herrings and extra numerals too, so that the game does not become too simple.

Number of players
The number of players needed depends upon the times table chosen. When working on smaller tables (2, 3, 4, 5), you could play twice for each number, swapping parts. When working on larger tables, all of the children will be more involved but should still have opportunities to repeat games, taking on different roles. In this game, the 2x table can be played with 15+ or more children, but the 12x table will need at least 25 players.

Extension/challenge
Lay out the numbers 0–9 on the floor randomly, so that children don't have any clues in front of them and have to remember which two- and three-digit numbers appear in the table being studied. Ensure that there are enough numeral cards to form all of the answers at the same time. If you only supply the cards you need, the game will become easier as numerals are used up and there will be only the correct answer left by the time the children reach 12x. To maintain the difficulty level, include a few red herrings or extra numerals, which will be left on the floor at the end of the game.

A few more to go

Learning in sequence

Activity
Tell the children which times table they are working on. For example: 7 times table. Ask them to start to build a wall or tower or model with (7) bricks or pieces and say the sum: 1 x 7 = 7. Next they should add (7) more bricks or pieces, count them and say the sum: 2 x 7 = 14. Then they should add (7) more bricks or pieces, count them and say the sum: 3 x 7 = 21, and so on to the last bricks or pieces: 12 x 7 = 84.

After this, they should take away (7) bricks or pieces, count those left and say the sum: 11 x 7 = 77 and continue to take away (7) in this way until they reach the first number again: 1 x 7 = 7.

Extension/challenge
Each player should take a random handful of bricks or pieces and put them together to build a group wall or tower or model. They must then count the bricks or pieces that they have used and decide whether that number appears as a multiple within the times table that they have been working on. If it does, they must say the sum. For example: 5 x 7 = 35. If it doesn't, they must decide how many more bricks or pieces to add to make the nearest multiple, do so and then say the sum. For example: 39 bricks + 3 more bricks = 42. Now we have made 6 x 7 = 42.

Learning objective
Discovering the multiples within one complete times table in ascending and descending order sequence.

Preparation
Construction kits with at least 144 bricks or pieces each for one or more groups..

Number of players
Any number of players can take part in this game. Ask children to work in one or more groups of 2–6 players.

© Debbie Chalmers and Brilliant Publications Fun Games and Activities for Teaching Times Tables
This page may be photocopied for use by the purchasing institution only.

A model experience

Guessing the table

Activity
Secretly decide on a times table and give the children three multiple numbers that occur as answers within that table.
For example: 18, 24 and 72 (6 times table).

Children make three separate models (walls, towers or creative shapes) using each of the three numbers of pieces. They then guess which table has been chosen, by breaking their models down and separating them into equal pieces until they find the one factor that they all have in common. If they struggle to find the answer, you could choose to offer another multiple number to make it more obvious.

Encourage the children not to just try each number, beginning with 2s, then 3s and so on, but to make a sensible guess and then check their answer with the bricks. Remind them that all even numbers will divide by 2, but that if they break any of their models down into more than twelve separate pieces it is not the 2 times table that has been chosen. Choose numbers randomly, rather than starting with the 2 times table, then the 3 times table and so on.

Learning objective
Understanding that each times table consists of a particular number multiplied over and over again.

Preparation
Construction kit containing at least 144 small bricks or similar pieces.

Number of players
Any number of players may participate in this game. Children may play individually or in pairs or small teams.

Extension/challenge
Turn the activity into a competition by asking children to work in teams and try to discover the correct factor first, from four clues that are given one at a time. The first two numbers should be those that appear in several different times tables, so that the children may narrow it down to two or three options, but the third and fourth numbers should confirm which of the options is the correct one.
For example: 12, 48, 54, 66
- 12 could be from the 2, 3, 4, 6 or 12 times table
- 12 and 48 could be from the 4, 6 or 12 times table
- 54 could be from 6 or 9 times table, (but the 9 times table does not contain 12 or 48)
- 66 confirms that it is the 6 times table.

Some children may guess correctly after the third clue; they all should guess after the fourth clue.

How many in the group?

Guessing the table

Activity
Ask 12 children to form a group, without cards. The rest of the children should sort the 0–9 cards into piles. If there are ten children, or ten pairs of children, they may hold one set of 13 identical cards each; if there are fewer children, they will each take charge of more than one number.

Call out the numbers that are the answers to the questions of a particular times table (missing out the 1x answer, which would give away the table immediately), beginning with 2x. The children with the 0–9 cards form the numbers by standing together and holding up the numerals in the correct orders. They can place each set on the floor as they move on to create the next number.

Learning objective
Visualising the multiples within a particular times table in ascending order.

Preparation
13 sets of number cards 0–9.

Number of players
Ideally 14–22 children but, if the group is larger, they may work in pairs to use the 0–9 cards.

The group of 12 children discusses the numbers and decides which times table they think is being described. That number of children then remain standing and the others sit. For example: if the numbers called are 16, 24, 32, 40…, then it is the 8 times table and 8 children stand and 4 sit.

Once the children are familiar with the game, you can introduce an element of competition by speaking the numbers more quickly and encouraging them to guess within as few answers as possible. Needing to move quickly to form the numbers and to stand up and sit down makes the game more fun for all players.

Extension/challenge
Instead of beginning with the 2x answer and working in ascending order sequence, begin with 12x and give the answers in descending order sequence.

For more able or experienced children, try giving out the answers in a random order, beginning with those which occur in several different tables and suddenly including one that gives the answer away.
For example: 24, 48, 72, 144
- 2, 3, 4, 6, 8 and 12 are considered
- 2 and 3 are eliminated, but 4, 6, 8 and 12 are still possible
- 4 is eliminated, but 6, 8 and 12 are still possibilities
- 6 and 8 are no longer considered, as this number only occurs in the 12 times table so 12 children stand.

Pick me up

Guessing the table

Activity
Tell the children which times table they are studying and then turn over two cards. If possible, put the two cards together and lay them out for the group, face up. If not possible, swap one or both cards to form a multiple answer.

A child turns over two cards and either makes another two-digit number, or uses the cards separately to make two single digit numbers, or adds a card to one of the two-digit numbers to make a three-digit number. The next child turns over two more and either makes another one or two numbers, starts a new number or adds the third digit to a number already made. Turns continue in this way until there are no cards left face down.

If a whole times table is completed, choose another times table and take two cards again to start it for the children, who then take turns to add to it as before.

Extension/challenge
Ask children to lay out the numbers that they form in order, filling in the gaps as they find the right numerals.

More able children, working as an extension group, may enjoy the challenge of completing two or more tables simultaneously.

Learning objective
Recognising all of the multiples within one times table, when presented in any order.

Preparation
13 sets of number cards 0–9. Lay all of the cards face down, and spread them out in a random pattern.

Number of players
Any number of players may participate in this game. If there are 12 players or fewer, one table may be enough in one session, but more players will need to work on more tables, one after another.

Find your partners

Guessing the table

Activity
Divide the children into two teams, called tens and units, and give them some form of identification, such as coloured bands, tags, bibs or sashes. Hand each player an appropriate number card. For example, for the 8 times table:

Units team	8	6	4	2	0	8	6	4	2	0	8	6
Tens team	0	1	2	3	4	4	5	6	7	8	8	9

Ask them to walk around the space, talking to each other and each finding their partner from the other team, until they have created all of the multiples of a particular times table and named it.

Explain to the children that the unit numerals often repeat more than the tens numerals. (The 5 times table is particularly easy, as the units are all 0 and 5, so this may be a good choice to introduce and practise the game. Move on to a harder one once all children understand how to play.)

If some pairs form multiples from a different table, but then find that the rest of them cannot be made from the numerals that are left, they will need to split up again and find other partners. The game continues until all twelve pairs are formed correctly and the times table agreed by all players.

Learning objective
Creating all of the multiples that occur within a particular times table.

Preparation
2 sets of cards that form the multiples of one particular times table, divided into the separate digits of the tens and the units. (Only use the 2–8 times tables for this game, as the 9–12 times tables include some three-digit multiples and involving hundreds makes it harder.)

Number of players
24 players (if there are not enough children, adults should join in)

Extension/challenge
Play the game with the 9–12 times tables instead, including the hundreds with the tens where needed, as 10, 11, 12, 13, 14.

The 9 times table is quite easy as almost all of the numerals are different. The 10 times table is very easy and obvious, as soon as the children realise that all of the units are 0. The 11 times table is easy once the children realise that most of the numerals in the two groups are the same. However, the 12 times table can be confusing at first, as the hundreds and tens numerals descend but miss out 11 and 5, while the units numerals repeat in blocks of five.

The challenge set for the players could be simply to work out as quickly as possible which table they are working with and then to form the pairs rapidly.

Show me your answers

Activity

Divide the children into groups of 12, without cards. If the class or group divides by 12 exactly, you will need to hold up the cards. If not, a few children can take charge of the numeral cards.

Call out a number that appears as the answer within the 2–12 times tables. Whoever has the cards 0–9 forms the one-, two- or three-digit number by holding up the numerals in the correct order for the everyone to see.

Learning objective
Linking the factors and multiples that occur within each times table.

Preparation
2 sets of number cards 0–9.

Number of players
This game needs at least 24 players and is more fun with 36 or more.

Each group of 12 children decides as quickly as possible on a times table that includes the number as a multiple answer and moves so that the correct number of children are standing and the others are sitting down. It may be necessary to emphasise that it is the answers, not the questions, that they must think about. If the number 8 was called out, for example, a standing group of 2 or 4 would be correct but a group of 8 would not.

Some number answers occur in several different times tables, while some appear in only one. Impress upon the children that they should not look at other groups or try to copy each other, as groups may sometimes choose different solutions and be equally correct. For example:

| 49 | 7 children stand and 5 sit |
| 30 | 3, 5, 6 or 10 children stand and the others sit |

When groups choose different correct answers to the same solution, discuss which numbers are paired as factors of the larger number. For example:

| 30 | 3 and 10 are factors and so are 5 and 6 |

Extension/challenge

Invite two groups of 12 players each to secretly choose a number 1–12 and to represent it by standing and sitting as in the game above. Then ask them to decide which multiple number will be formed if the two groups are multiplied by each other. Each group fetches the appropriate numeral cards to make the answer and shows the number to the other group. For example: If 6 children are standing in one group and 3 children are standing in the other group, they will need to fetch the numerals 1 and 8 and show 18 to each other.

One, two, three

Guessing the table

Activity
Choose a times table to work on and give the children clues by telling them one number at a time that occurs as a multiple answer within it. Aim not to give away the table too quickly, by choosing mostly numbers that occur in more than one table and slipping in a bigger clue at intervals. For example: This table has 36 in it. It has 72, and 45. It has 108 and 18. It has 27 and 9... .

Children take the correct numerals from the piles and lay out the numbers as they hear them, under the H, T and U (Hundreds, Tens and Units) headings, discussing the numbers with each other if they wish. Children should call out the times table being described as soon as they are sure which one it is, aiming to beat their own score and guess the tables after fewer and fewer clues.

Learning objective
Understanding which times tables can be used when working with particular numbers.

Preparation
- 13 sets of number cards 0–9, sorting into piles by number
- 3 larger cards showing H, T and U, laid out on the floor.

Number of players
Any number of players.

You will easily be able to observe which children are most confident and able to lead this game and which have the strongest knowledge of the numbers within different tables and are able to call them to mind and use them most quickly. Those who are less confident and secure in their knowledge will benefit from practising and working with the others.

Extension/challenge
Leave out the single- and three-digit numbers, as they often give away the answer. Tell children only two-digit numbers and let them put the numerals together and lay them out in spaces on the floor at random. Encourage children to guess more and more quickly, as they become more comfortable working with the range of numbers.

Eventually, very able children working as an extension group may dispense with the numeral cards and try to picture the numbers in their heads and remember them while playing the game.

Look underneath

Guessing the table

Activity
Children take turns to roll the balls and knock down the skittles, then lift them up and remember their numbers. When they have knocked down enough skittles and seen enough numbers to recognise which table they are representing, they shout out the answer.

Arrange the skittles carefully, so that the easiest to knock down are the ones with the more commonly used multiples under them (eg, for the 8 times table,16, 24, 40, 80). Ensure that the ones that give the table away quickly are further away and harder to knock down (64, 56, 96).

Extension/challenge
Use 22 skittles and the multiples of two different times tables. Tape the numbers under the skittles at random, so that the two tables are mixed together. (Do not include the 1x numerals.)

Allow the children to roll balls and guess as before, but ask them not to speak until they are prepared to try to name both times tables.

Learning objective
Identifying a times table using knowledge of its multiples as clues.

Preparation
- 11 skittles. Write the multiples of a particular times table onto pieces of paper and tape one to the bottom of each skittle, so that they cannot be seen while the skittles are standing up. (Do not include the 1x numeral as this would give the answer away.)
- Balls that can be rolled to knock down the skittles.

Number of players
Any number of players. Players cooperate and work as a whole group.

Fun Games and Activities for Teaching Times Tables © Debbie Chalmers and Brilliant Publications

Just choose one

Guessing the table

Activity
Divide the children into two teams of two or more players. Ask each team to stand beside their bucket with a pile of 12 beanbags.

Call out a multiple from a times table and the children in each team discuss and try to guess the table. When they agree on their first guess, they throw that many beanbags into their bucket.

Call out another multiple from the same times table and the children decide together whether they still think that their guess is correct. They may choose to keep the same number of beanbags in the bucket or to change their minds and add more or take some out.

Learning objective
Identifying a times table using knowledge of its multiples as clues.

Preparation
- 24 beanbags
- 2 buckets, boxes or crates.

Number of players
Four or more children can play this game, but it is fairest if there is an even number so that each team has the same number of players.

Continue with the game in this way until one or both teams are sure that they are correct and choose to tip over their bucket to count their beanbags and announce their decision.

The 5, 10 and 11 times tables are very easy to guess quickly and so can be good early choices for building confidence and understanding of the game. Gradually move on to choosing the ones that are the least obvious because they have many multiples similar to those in other tables, such as the 4, 6 and 8 times tables.

Extension/challenge
Introduce gentle competition. If enough beanbags and buckets are available (or other similar resources), extend the game to four or more teams.

Call out only three multiples and then ask each team to settle on a decision or a guess. They should all tip over their buckets and show their answers at the same time, to see whether they agree and, if not, who has found the correct answer.

Make a record

Guessing the table

Activity
Divide the children into two groups with equal numbers of players – Team A and Team B.

Give balls to Team A and the mini-whiteboards or paper and mark making tools to Team B. You could supply one ball or set of mark making tools per pupil or they could be asked to share.

Quietly tell Team A a multiple number that occurs within a times table. Two players from Team A bounce balls to represent the number. For example, if the number is 64, the first player bounces 6 times and the second player bounces 4 times.

Team B listens to the bounces and counts carefully and records the numbers heard by writing down the figure that they make.

Learning objective
Identifying a times table using knowledge of its multiples as clues.

Preparation
- ❏ 2 or more balls (foam ball indoors; football outdoors)
- ❏ Mini-whiteboards and marker pens, or sheets of A5 paper and pencils.

Number of players
Four or more players may participate in this game. An even number is required.

Repeat with more multiples of the same times table (eg 16, 24, 88), until Team B guesses the times table represented by Team A's bounces. (Answer 8.) If Team B cannot guess, ask Team A if they know the answer. Then swap roles and repeat the game.

Extension/challenge
Give both teams balls and whiteboards or paper. Quietly tell each team in turn multiples of a different table, while they take turns to bounce balls and record the numbers they hear, competing to guess the other team's times table first.

Follow the trail

Guessing the table

Activity
Call out a series of numerical clues to a particular times table and invite children to follow the trail to find the correct answer. Children discuss the clues and turn over their cards to eliminate numbers, until one is left to be the correct answer. Clues should be simple at first and become gradually harder, requiring more knowledge of times tables.

For example:

- 3x this number is more than 10 (turn over 2 and 3)
- 7x this number is less than 70 (turn over 10, 11 and 12)
- 4x this number is more than 25 (turn over 4, 5 and 6)
- 9x this number is more than 66 and less than 79 (turn over 7 and 9).
- Answer is 8.

Learning objective
Identifying a times table using knowledge of its sums and some creative thinking.

Preparation
Provide two or more sets of number cards 2–12 (one set for each team of players). Lay out a set of number cards for each team.

Number of players
Four or more children may participate in this game, forming teams of two or more players.

Extension/challenge
Be more creative with clues, so that children need a more confident working knowledge of the multiple figures that occur within the times tables in order to picture which numbers fit the descriptions.

For example:
- 4x this number has 2 digits and 1 is half as big as the other (36)
- This number x itself is just 1 more than a number of tens (81)
- Answer is 9.

© Debbie Chalmers and Brilliant Publications   Fun Games and Activities for Teaching Times Tables
This page may be photocopied for use by the purchasing institution only.

Swap with me

Finding factors

Activity
Give two cards to each child and ask them to hold them up so that they are visible to the other players, one in each hand.

Children walk around with their number cards. Point to a child and tell them a number that is a multiple answer within a times table. The child thinks of two factors that can make up the multiple and goes to other children who are holding those numbers, asking to swap for the numbers they have.

For example: a child holding 4 and 5 is given the multiple number 88 and must swap one card with a child holding 8 and the other card with a child holding 11.

Learning objective
Recognising different numbers that can be factors of the multiples that appear within the times tables.

Preparation
2 sets of number cards 2–12.

Number of players
Ideally, play this game with 22 players, but some extra cards showing the most popular numerals can be included if there are more players.

When the cards have been swapped and the child has shown you the correct factors, it is another player's turn.

Try not to choose numbers with factors that are on the cards the child is already holding, so that each turn involves two swaps. Many numbers will have more than one possible pair of factors and each child may choose whichever they like or can think of first. If a child is stuck, you could ask another player to help out by giving some clues.

Extension/challenge
Give each child four cards and choose multiples with two or more sets of factors. Ask children to find the two sets by swapping four cards with other players during their turns. It may not be possible to ensure that four swaps are needed for every turn, but players must identify for themselves which cards to swap and which to keep.

For example: a child holding 3, 6, 7 and 10 is given the number 72, so must keep the 6 and swap the 3, 7, and 10 for 12, 8 and 9.

Factor that in

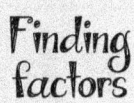
Finding factors

Activity
Give out two identical cards 2–12 to 11 children, so that they each represent a different numeral, and ask them to sit in spaces with one numeral in their hands and one on the floor beside them.

Use one, two or three of the 0–9 cards at a time to hold up different numbers that occur as multiple answers within times tables 2–12 and call out each number aloud in turn for the group.

Children stand up and find their partners and walk to you together if they are factors of the multiple. This may involve one or more pairs of children. For example:

| 20 = | 2 and 10; 4 and 5 |
| 21 = | 7 and 3 only |

Learning objective
Linking factors and multiples that work together.

Preparation
- ❏ 2 sets of number cards 2–12
- ❏ 2 sets of number cards 0–9.

Number of players
This game can be played with 11 or more children, as some or all numerals can be represented by single players or pairs or small groups.

If the factors of a number are a numeral squared, the child picks up their spare numeral from the floor and walks to you alone, holding up both numerals. For example:

| 4 = | 2 and 2 |
| 49 = | 7 and 7 |

Some numbers require children to think of both types of factor and three sets altogether. For example:

| 36 = | 3 and 12; 4 and 9; 6 and 6 |

You must ensure that all children are included often enough. It is a good idea to plan and write a list of the numbers to call out in advance, until you are experienced in leading this game. You may wish to call out 24 as the last number, as it involves six children as factors and brings the game to a satisfying end.

Extension/challenge
Hold up the numbers that are the multiple answers to a particular times table, one after another in sequence, but miss one out and ask children to identify the table and the missing number, so that the two factors can walk to you. This is a team building, group work exercise in which children have to think, discuss and call out to each other.

© Debbie Chalmers and Brilliant Publications Fun Games and Activities for Teaching Times Tables
This page may be photocopied for use by the purchasing institution only.

Either way

Finding factors

Activity
Give 20 children one 0–9 card each. Divide the rest of the children into two groups of two or more. Give cards 1–12 to one group and cards 2–12 to the other group.

Call out a number that is the multiple answer to one or more sums within particular times tables. The children with the 0–9 cards form the number by standing in the centre of the space, separately or in pairs or in lines of three, and holding up the number cards in the correct orders.

Learning objective
Linking numbers that are factors and multiples of each other.

Preparation
- Set of number cards 1–12
- Set of number cards 2–12
- 2 sets of number cards 0–9

Number of players
24+ players

The children in the two groups confer and decide how each answer can be made and in how many different ways. They then take those numerals to the centre, in pairs, one from each group, to join those already there. For example:
- 72 is called, so two sets of children holding up 7 and 2 walk to centre and stand together.
- 6 x 12 and 12 x 6, 8 x 9 and 9 x 8 are decided upon, so a child holding 6 and a child holding 8 from one group
- and a child holding 12 and a child holding 9 from the other group walk to centre to join them.

Reinforce that multiplication works in the same way whichever order the numbers are used by asking the pairs to move around each other and chant the phrase twice each, to prove that either order produces the same answer. For example:
- 6 x 12 is 72;   12 x 6 is 72   and
- 8 x 9 is 72;   9 x 8 is 72

All players then return to their groups, ready to work on the next number called.

Extension/challenge
Ask children to guess which multiple occurs most frequently within times tables 2–12. They may hit upon the answer by lucky guess or spend a long time discussing within groups and working it out. Some classes may take it up as an interesting challenge to work out and discover with their families at home. (Numbers 12, 18, 20, 30, 40, 48, 60 and 72 each occur four times within the tables and 36 occurs five times, but 24 is the most frequent, occurring six times, in tables 2–12.)

It works both ways

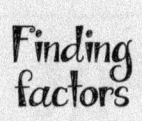
Finding factors

Activity
Ask 12 children each to hold two identical number cards 1–12 and stand in a line beside each other. Shuffle and place all the 0–9 cards face down in a pile and ask the rest of the children to form pairs and stand in a line behind each other, next to the cards. The first pair of children turns over the top two cards from the pile and holds them up, maintaining the order in which they were turned.

Learning objective
Visualising the range of factors for particular multiples.

Preparation
- 2 sets of number cards 1–12
- 13 sets of number cards 0–9.

Number of players
14+ players. An even number of players is needed.

The pair decides together upon two factors that can be multiplied to make that number and go to the line to bring those children out to stand together, each holding up one numeral. If they decide to use one numeral squared, they will bring out only one child and ask that child to hold up both of their numerals, one in each hand. For example:

Number cards	0 4	1 5	2 8	3 0	4 9	5 5	6 4	7 2	8 4	9 0	9 6
Factor cards	1 & 4	3 & 5	4 & 7	6 & 5	7 & 7	11 & 5	8 & 8	9 & 8	12 & 7	9 & 10	8 & 12

The whole group check that the factors are correct and chant the phrase. For example: 8 x 12 is 96; 12 x 8 is 96.

Those holding answer numerals return to their line, the two children sit down with their factor cards and the next two turn over the top two cards from the pile to take their turn.

Occasionally, cards will be taken that do not make a number that appears in the times tables. In these cases, you can give permission for the cards' order to be swapped. For example: 61 can become 16 or 52 can become 25. If this still does not produce a usable answer, you could ask for the cards to be randomly mixed back into the pile and shuffled and the children are then allowed to take two new cards. For example: neither 34 nor 43 has factors within the times tables.

Extension/challenge
Remove some cards from the pile, keeping only 0, 1, 2, 3, 4 and 8. Ask the children to form groups of three instead of pairs and to take turns to pick up three cards from the top of the pile instead of two, then to move the order of the cards around until they can create a three-digit multiple answer for which they can find factors. You will discover which children know or work out that they only need to consider numerals 9–12 as factors for the three–digit multiples.

Put them together

Finding factors

Activity
Children take turns to play, rolling two balls to knock down as many skittles as they can.

Players knock down skittles and check their numbers. They must then think of as many times table phrases as they can involving those numbers and score a point for each correct question and answer. For example: if a child knocks down skittles 5, 3 and 8, they may think of 5 x 3 = 15 and 3 x 8 = 24 and 5 x 8 = 40 and score 3 points.

No extra point is scored for giving the inverse relation (for example: 5 x 3 = 15 and 3 x 5 = 15). If a child cannot multiply the skittles they knock down and give the correct answer, they do not score the point.

Learning objective
Identifying the multiples that are made by combining different factors.

Preparation
- 12 skittles labelled with numbers 1-12, set up in random order
- 2 balls of suitable size and weight
- line marked on the ground behind which children must stand or kneel to roll the balls.

Number of players
Any number of players can take part in this game.

A child may score points by having exceptionally good aim, and so knocking down the most skittles or the numbers which are easiest to multiply, or by having an exceptionally good knowledge of times tables and so being able to supply the correct answers no matter which they knock over. With practice, children will improve both skills.

Encourage children to aim to achieve their own best scores, rather than to compete with each other. Reassure them that, as in most games, there is an element of luck involved, since some players may create easier sums to work with by knocking down the skittles 1, 2 and 10, while others may make more difficult sums because they knock down the skittles 6, 9 and 12.

Extension/challenge
Place the skittles with numbers such as 2, 5, and 10 further away or at more difficult angles, so that children are more likely to knock over the harder numbers to work with, such as 7, 8 and 12.

Remove the number 1 skittle, so that no points can be scored by reciting 1x any other number.

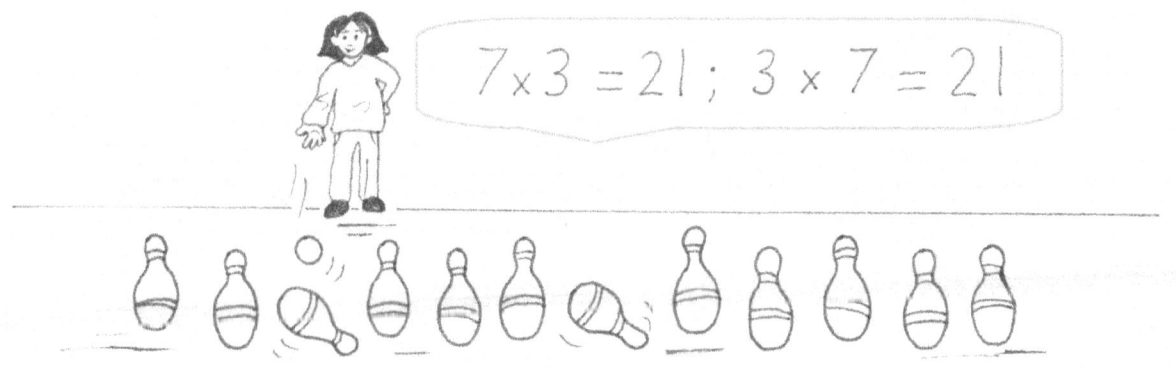

Fun Games and Activities for Teaching Times Tables   © Debbie Chalmers and Brilliant Publications
This page may be photocopied for use by the purchasing institution only.

Two bounces

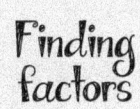
Finding factors

Activity
Divide children into pairs and give each pair a ball. Call out a multiple number and the children must confer with each other and bounce their ball to represent two numbers that can be factors of that multiple. For example: 35 is called, so one child bounces the ball 7 times and the other bounces it 5 times.

Ask each pair of children to stand up in turn, then call out a different multiple number and allow them a little time to discuss and decide their answer. Ask the other players, who are seated, to be thinking of an answer too, in case it is needed. If a pair cannot think of two factors, or if the ones they choose are not correct, allow another pair to demonstrate their answer and then give the first pair a second turn with a different (simpler) multiple, so that they still experience success and do not lose confidence.

Learning objective
Identifying the different numbers that can be factors of the multiples that appear within the times tables.

Preparation
1 ball for each pair of children, large enough for the children to throw, bounce and catch easily (foam ball indoors, football outdoors).

Number of players
Any even number of players may participate in this game.

If the chosen multiple has more than one pair of factors, accept the first answer and then ask the other pairs whether any of them have thought of a different answer that they would like to demonstrate.

When the multiple is a squared number (for example 16) and the players bounce their ball the same number of times (4 bounces each), all players should stand up and throw the balls to each other until each pair has swapped with another. This ensures that all players participate in thought and movement at frequent random intervals and do not lose concentration or stop thinking about the game. Call out a squared number whenever some children appear to be losing focus, to bring the group back together.

Extension/challenge
Allow each pair of players to secretly choose their own multiple and then bounce their ball to indicate two factors so that the rest of the group can guess.

Aim to pair children of similar abilities, to avoid one dominating the game while the other follows instructions and does not learn, or one becoming frustrated when the other can only manage simpler factors and multiples.

Throw them to us

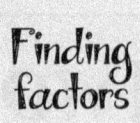

Finding factors

Activity
Ask the children to stand in two groups. There must be at least 12 players in each group and one player left over to take the first turn.

Place the beanbags in a pile between the two groups, choose one child at a time to stand beside them and then call out a multiple number. The chosen player must think of two factors that go together to make the multiple and throw beanbags to the right number of children in each group. For example: If 55 is called, beanbags are thrown to 5 children in one group and 11 children in the other group.

The children holding the beanbags then walk out of their groups and hold them up to illustrate the sum they are representing, while the first player says the sum aloud for the group to hear. For example: 55 = 5 x 11.

After this, the children replace the beanbags on the centre pile and rejoin their groups. Another player is chosen to make the next sum and swaps places with the player in the centre.

The game should be repeated until each player has taken a turn to throw the beanbags, but this may need to happen during two or more games on separate days.

Learning objective
Recognising the different numbers that can be factors of the multiples that appear within the times tables.

Preparation
- 24 beanbags
- List of assorted multiple numbers that occur within the times tables, in a random order (one number for each player).

Number of players
25+ players.

Extension/challenge
Work with small groups at a time and encourage children to work together to collect beanbags and create the sums for multiples that have two or three sets of factors. Unless a very large collection of beanbags is available, it will be necessary to make one sum at a time, show it to you and then re-use the beanbags to make the next sum. For example: 36 = 12 x 3. 36 = 9 x 4. 36 = 6 x 6.

The nine multiples that have two sets of factors are: 12, 16, 18, 20, 30, 40, 48, 60 and 72. There are two multiples that have three sets of factors: 24 and 36.

Two bounces

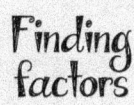

Finding factors

Activity
Divide children into pairs and give each pair a ball. Call out a multiple number and the children must confer with each other and bounce their ball to represent two numbers that can be factors of that multiple. For example: 35 is called, so one child bounces the ball 7 times and the other bounces it 5 times.

Ask each pair of children to stand up in turn, then call out a different multiple number and allow them a little time to discuss and decide their answer. Ask the other players, who are seated, to be thinking of an answer too, in case it is needed. If a pair cannot think of two factors, or if the ones they choose are not correct, allow another pair to demonstrate their answer and then give the first pair a second turn with a different (simpler) multiple, so that they still experience success and do not lose confidence.

Learning objective
Identifying the different numbers that can be factors of the multiples that appear within the times tables.

Preparation
1 ball for each pair of children, large enough for the children to throw, bounce and catch easily (foam ball indoors, football outdoors).

Number of players
Any even number of players may participate in this game.

If the chosen multiple has more than one pair of factors, accept the first answer and then ask the other pairs whether any of them have thought of a different answer that they would like to demonstrate.

When the multiple is a squared number (for example 16) and the players bounce their ball the same number of times (4 bounces each), all players should stand up and throw the balls to each other until each pair has swapped with another. This ensures that all players participate in thought and movement at frequent random intervals and do not lose concentration or stop thinking about the game. Call out a squared number whenever some children appear to be losing focus, to bring the group back together.

Extension/challenge
Allow each pair of players to secretly choose their own multiple and then bounce their ball to indicate two factors so that the rest of the group can guess.

Aim to pair children of similar abilities, to avoid one dominating the game while the other follows instructions and does not learn, or one becoming frustrated when the other can only manage simpler factors and multiples.

Throw them to us

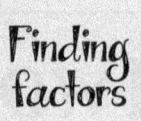
Finding factors

Activity
Ask the children to stand in two groups. There must be at least 12 players in each group and one player left over to take the first turn.

Place the beanbags in a pile between the two groups, choose one child at a time to stand beside them and then call out a multiple number. The chosen player must think of two factors that go together to make the multiple and throw beanbags to the right number of children in each group. For example: If 55 is called, beanbags are thrown to 5 children in one group and 11 children in the other group.

The children holding the beanbags then walk out of their groups and hold them up to illustrate the sum they are representing, while the first player says the sum aloud for the group to hear. For example: $55 = 5 \times 11$.

After this, the children replace the beanbags on the centre pile and rejoin their groups. Another player is chosen to make the next sum and swaps places with the player in the centre.

The game should be repeated until each player has taken a turn to throw the beanbags, but this may need to happen during two or more games on separate days.

Learning objective
Recognising the different numbers that can be factors of the multiples that appear within the times tables.

Preparation
- 24 beanbags
- List of assorted multiple numbers that occur within the times tables, in a random order (one number for each player).

Number of players
25+ players.

Extension/challenge
Work with small groups at a time and encourage children to work together to collect beanbags and create the sums for multiples that have two or three sets of factors. Unless a very large collection of beanbags is available, it will be necessary to make one sum at a time, show it to you and then re-use the beanbags to make the next sum. For example: $36 = 12 \times 3$. $36 = 9 \times 4$. $36 = 6 \times 6$.

The nine multiples that have two sets of factors are: 12, 16, 18, 20, 30, 40, 48, 60 and 72. There are two multiples that have three sets of factors: 24 and 36.

Walking two by two

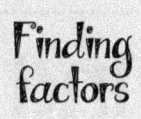

Finding factors

Activity
Ask children to stand in two lines, one behind each number line. Call out a multiple number and the two children at the front of the lines must confer with each other to choose two factors that go together to make that multiple and who will stand by which. They then walk to those places for the rest of the group to check their answer.

If they are not correct, others in the lines may call out suggestions and the players may walk to other numbers until they have found two factors that are right. They then return to rejoin their lines at the back.

Learning objective
Recognising the different numbers that can be factors of the multiples that appear within the times tables.

Preparation
2 sets of number cards 2–12, laid out as parallel number lines, a short distance apart.

Number of players
Any even number of players may participate in this game.

The game continues with a new multiple called for each pair until all players have taken a turn. If the group is small, they could each take several turns but, if it is large, one or two turns each may be enough.

Extension/challenge
Ask the children to play the game without discussing the factors with each other. When the multiple number has been called, the child at the front of Line A should walk along to stand beside one of the factors. The child at the front of Line B must then walk along to stand beside the number which is the matching factor. The two factors must work together. For example: If 20 is called and the first player chooses 4 as a factor, the second player must choose 5, not 2 or 10. When the group agrees that the factors are correct, the two players rejoin the backs of their lines to wait for another turn.

In a mixed ability group, choose as many multiples as possible that have more than one set of factors. Aim to put less confident children into Line A, so that they may choose first and use any factors that they know or can guess. Put more able children into Line B, so that they will be more challenged by having to match specific factors.

Find a partner

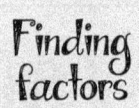
Finding factors

Activity
Give out a number card to each player and ask them to stand in a line, holding up their cards in order 1–12. The number 1 should be held by the least able child (or by an adult) in each line.

Call out a multiple and the children representing the factors of that number walk out of their line and forward to each other, shake hands and then return to their places. Players can walk at the same time as each other or one after another, as they think of the correct answers. If one child knows the pair of factors but the child holding the other card does not, they can walk right up to them and lead them out of the line to shake hands. For example:

Learning objective
Understanding the different numbers that can be factors of the same multiples that appear within the times tables.

Preparation
Set of number cards 1–12.

Number of players
This game is suitable for 11–12 players working as one group.

8 =	1 and 8, 2 and 4
12 =	1 and 12, 2 and 6, 3 and 4
24 =	2 and 12, 3 and 8, 4 and 6
30 =	3 and 10, 5 and 6

Only the numbers 1–12 are possible factors, so the number 1 may only be a factor for any multiple up to 12, the number 2 may only be a factor for any multiple up to 24, and so on.

You may or may not wish to include multiples that only have the factors 1 and themselves (for example, 1, 2, 3, 5, 7 and 11). Avoid using squared numbers as multiples in this game (4, 9, 16, 25, 36, 49, 64, 81, 100, 121, 144), as there will not be two players representing the same number within the line.

Extension/challenge
Play with two groups together. There will need to be 24 children or 22 children and two adults as players.

Ask the players to stand in two lines, facing each other, holding up their number cards 1–12. They will notice that making the numbers read in order from right to left means that the number 1 is opposite the number 12. When each multiple number is called, the factors walk across the centre space to shake hands with their partners from the other line. Squared numbers may now be included as multiples, as identical numbers will exist in the two lines.

Set it out

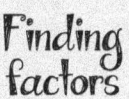

Finding factors

Activity

Divide the children into pairs. Give each pair 3 cards, which show the x and = symbols and a multiple number, and read the number aloud with them to ensure that they understand it correctly.

The children decide on the factors for their multiple number and go to the two trays. The first player takes the right number of straws to represent one factor from one tray and the second player takes the right number of straws to represent the other factor from the other tray.

They then work together to lay out their sum on the ground or floor, by placing the correct number of straws, then the x symbol, then the other straws, then the = symbol and then the multiple number in a line reading from left to right.

Learning objective
Creating the sums that occur within the times tables.

Preparation
- ❏ Number cards showing a selection of multiples that occur within the times tables. Allow at least two different numbers for each pair of children.
- ❏ Cards with symbols x and = (one set for each pair)
- ❏ 2 trays with large collection of drinking straws on each.

Number of players
Any even number of players may participate in this game.

For example: a pair is given x and = and 63. One child takes 7 straws and the other child takes 9 straws. They make 7 (straws) x 9 (straws) = 63.

Each pair shows and reads out their sum to the group in turn.

Ask the children to replace all the straws on the trays, then offer another multiple number card to each pair and repeat the game. This can continue until the cards run out. The number of turns may depend upon the size of the group.

Extension/challenge

Call out multiples that have two sets of factors and ask the children to work in their pairs to find both. They should then lay out their straws and move their symbol and number cards to explain both their answers to the group. Warn them that the inverse relation does not count as a second answer. For example:
   6 x 12 = 72 and 8 x 9 = 72 is correct.
   6 x 12 = 72 and 12 x 6 = 72 is not correct

The nine multiples that have two sets of factors are: 12, 16, 18, 20, 30, 40, 48, 60 and 72. There are also two multiples that have three sets of factors: 24 and 36. More able children could be challenged to work out the three separate sums for those multiples.

www.ingramcontent.com/pod-product-compliance
Lightning Source LLC
Chambersburg PA
CBHW081350160426
43197CB00015B/2723